P9-CFJ-165

Chicago Public Library
C
P
L
REFERENCE
Form 178 rev. 1-94

U·X·L COMPLETE LIFE SCIENCE RESOURCE

U·X·L COMPLETE LIFE SCIENCE RESOURCE

volume ONE: A-E

LEONARD C. BRUNO
JULIE CARNAGIE, EDITOR

AN IMPRINT OF THE GALE GROUP
DETROIT · SAN FRANCISCO · LONDON
BOSTON · WOODBRIDGE, CT

U•X•L Complete Life Science Resource

LEONARD C. BRUNO

Staff

Julie L. Carnagie, *U•X•L Senior Editor*
Carol DeKane Nagel, *U•X•L Managing Editor*
Meggin Condino, *Senior Market Analyst*
Margaret Chamberlain, *Permissions Specialist*
Randy Bassett, *Image Database Supervisor*
Robert Duncan, *Imaging Specialist*
Pamela A. Reed, *Image Coordinator*
Robyn V. Young, *Senior Image Editor*
Michelle DiMercurio, *Art Director*
Evi Seoud, *Assistant Manager, Composition Purchasing and Electronic Prepress*
Mary Beth Trimper, *Manager, Composition and Electronic Prepress*
Rita Wimberley, *Senior Buyer*
Dorothy Maki, *Manufacturing Manager*
GGS Information Services, Inc., *Typesetting*

Bruno, Leonard C.
U•X•L complete life science resource / Leonard C. Bruno; Julie L. Carnagie, editor.
p. cm.
Includes bibliographical references.
Contents: v. 1. A-E v. 2. F-N v. 3. O-Z.
ISBN 0-7876-4851-5 (set) ISBN 0-7876-4852-3 (vol. 1) ISBN 0-7876-4854-X (vol. 2)
1. Life sciences Juvenile literature. [1. Life sciences Encyclopedias.] I. Carnagie, Julie. II. Title.
QH309.2.B78 2001 00-56376

U•X•L, an Imprint of the Gale Group
27500 Drake Rd.
Farmington Hills, MI 48331-3535

Printed in the United States of America
10 9 8 7 6 5 4 3 2 1

Table of Contents

Contents

volume TWO: F-N

volume THREE: O-Z

Contents

Reader's Guide

U·X·L Complete Life Science Resource explores the fascinating world of the life sciences by providing readers with comprehensive and easy-to-use information. The three-volume set features 240 alphabetically arranged entries, which explain the theories, concepts, discoveries, and developments frequently studied by today's students, including: cells and simple organisms, diversity and adaptation, human body systems and life cycles, the human genome, plants, animals, and classification, populations and ecosystems, and reproduction and heredity.

The three-volume set includes a timeline of scientific discoveries, a "Further Information" section, and research and activity section. It also contains 180 black-and-white illustrations that help to bring the text to life, sidebars containing short biographies of scientists, a "Words to Know" section, and a cumulative index providing easy access to the subjects, theories, and people discussed throughout *U·X·L Complete Life Science Resource.*

Acknowledgments

Special thanks are due for the invaluable comments and suggestions provided by the *U·X·L Complete Life Science Resource* advisors:

- Don Curry, Science Teacher, Silverado High School, Las Vegas, Nevada
- Barbara Ibach, Librarian, Northville High School, Northville, Michigan
- Joel Jones, Branch Manager, Kansas City Public Library, Kansas City, Missouri
- Nina Levine, Media Specialist, Blue Mountain Middle School, Peekskill, New York

Comments and Suggestions

We welcome your comments on this work as well as your suggestions for topics to be featured in future editions of *U·X·L Complete Life Science Resource.* Please write: Editors, *U·X·L Complete Life Science Resource,* U·X·L, 27500 Drake Rd., Farmington Hills, MI 48331-3535; call toll-free: 1-800-877-4253; fax: 248-699-8097; or send e-mail via www.galegroup.com.

Introduction

U·X·L Complete Life Science Resource is organized and written in a manner to emphasize clarity and usefulness. Produced with grades seven through twelve in mind, it therefore reflects topics that are currently found in most textbooks on the life sciences. Most of these alphabetically arranged topics could be described as important concepts and theories in the life sciences. Other topics are more specific, but still important, subcategories or segments of a larger concept.

Life science is another, perhaps broader, term for biology. Both simply mean the scientific study of life. All of the essays included in *U·X·L Complete Life Science Resource* can be considered as variations on the simple theme that because something is alive it is very different from something that is not. In some way all of these essays explore and describe the many different aspects of what are considered to be the major characteristics or signs of life. Living things use energy and are organized in a certain way; they react, respond, grow, and develop; they change and adapt; they reproduce and they die. Despite this impressive list, the phenomenon that is called life is so complex, awe-inspiring, and even incomprehensible that our knowledge of it is really only just beginning.

This work is an attempt to provide students with simple explanations of what are obviously very complex ideas. The essays are intended to provide basic, introductory information. The chosen topics broadly cover all aspects of the life sciences. The biographical sidebars touch upon most of the major achievers and contributors in the life sciences and all relate in some way to a particular essay. Finally, the citations listed in the "For Further Information" section include not only materials that were used by the author as sources, but other books that the ambitious and curious student of the life sciences might wish to consult.

Timeline of Significant Developments in the Life Sciences

c. 50,000 B.C. *Homo sapiens sapiens* emerges as a conscious observer of nature.

c. 10,000 B.C. Humans begin the transition from hunting and gathering to settled agriculture, beginning the Neolithic Revolution.

c. 1800 B.C. Process of fermentation is first understood and controlled by the Egyptians.

c. 350 B.C. Greek philosopher Aristotle (384–322 B.C.) first attempts to classify animals, considers nature of reproduction and inheritance, and basically founds the science of biology.

A.D. 1543 Flemish anatomist Andreas Vesalius (1514–1564) publishes *Seven Books on the Construction of the Human Body* which corrects many misconceptions regarding the human body and founds modern anatomy.

1615 The modern study of animal metabolism is founded by Italian physician, Santorio Santorio (1561–1636), who publishes *De Statica Medicina* in which he is the first to apply measurement and physics to the study of processes within the human body.

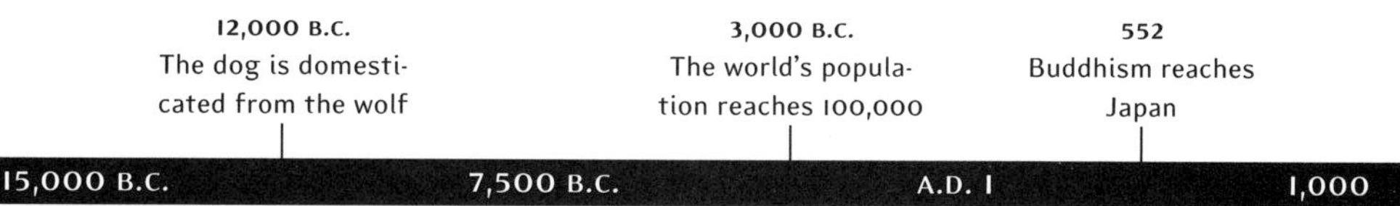

1628 The first accurate description of human blood circulation is offered by English physician William Harvey (1578–1657), who also founds modern physiology.

1665 English physicist Robert Hooke (1635–1703) coins the word "cell" and develops the first drawing of a cell after observing a sliver of cork under a microscope.

1669 Entomology, or the study of insects, is founded by Dutch naturalist Jan Swammerdam (1637–1680), who begins the first major study of insect microanatomy and classification.

1677 Dutch biologist and microscopist Anton van Leeuwenhoek (1632–1723) is the first to observe and describe spermatozoa (sperm). He later goes on to describe different types of bacteria and protozoa.

1727 English botanist Stephen Hales (1677–1761) studies plant nutrition and measures water absorbed by roots and released by leaves. He states that the plants convert something in the air into food, and that light is a necessary part of this process, which later becomes known as photosynthesis.

1735 Considered the father of modern taxonomy, Swedish botanist Carl Linnaeus (1707–1778) creates the first scientific system for classifying animals and plants. His system of binomial nomenclature establishes generic and specific names.

1779 Dutch physician Jan Ingenhousz (1739–1799) shows that carbon dioxide is taken in and oxygen is given off by plants during photosynthesis. He also states that sunlight is necessary for this process.

1802 The word "biology" is coined by French naturalist Jean-Baptiste Lamarck (1744–1829) to describe the new science of living things. He later proposes the first scientific, but flawed, theory of evolution.

1650 England's first coffee house opens

1710 The first copyright law is established in Britain

1770 The Boston Massacre occurs

1620 1680 1740 1800

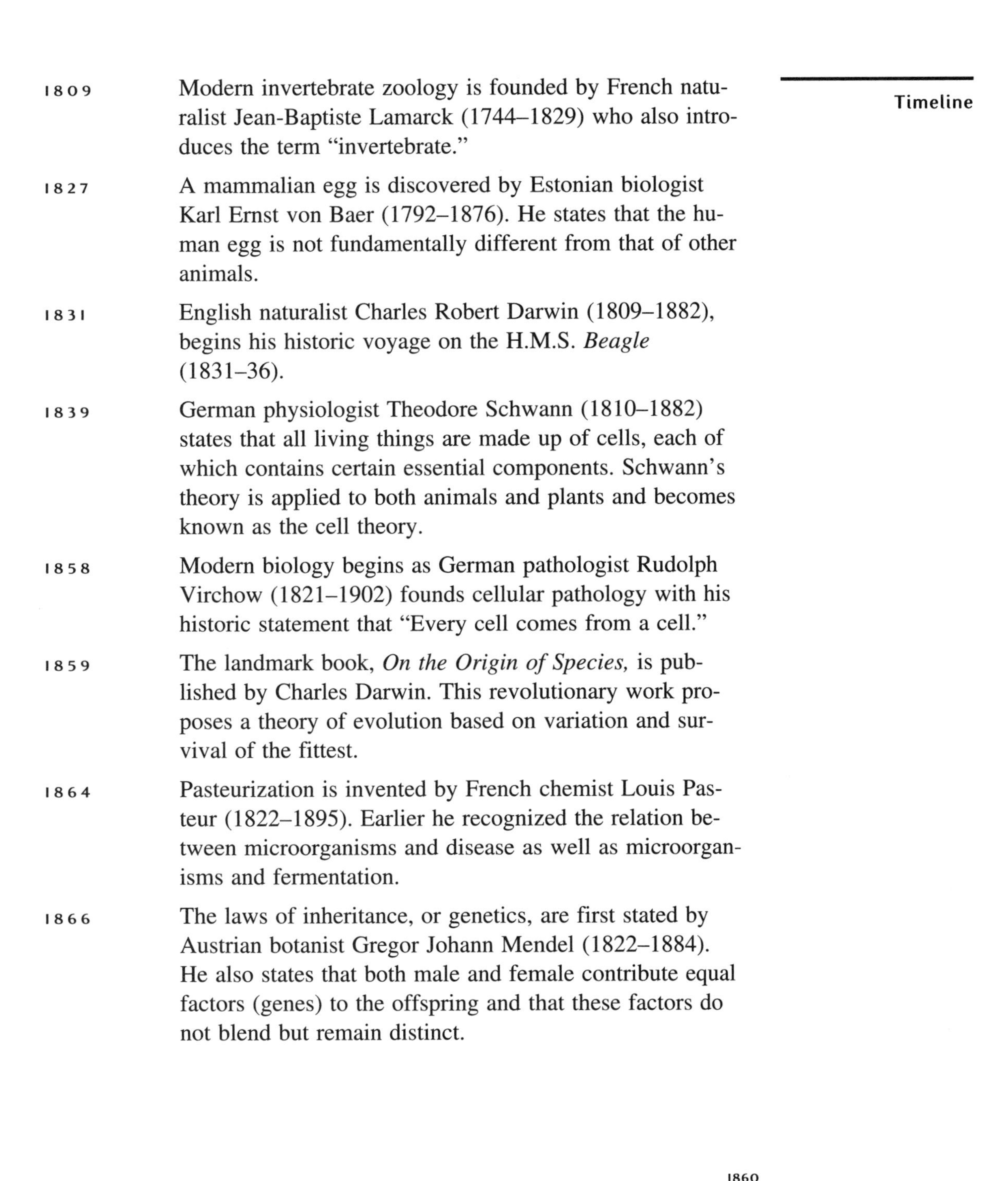

1809 Modern invertebrate zoology is founded by French naturalist Jean-Baptiste Lamarck (1744–1829) who also introduces the term "invertebrate."

1827 A mammalian egg is discovered by Estonian biologist Karl Ernst von Baer (1792–1876). He states that the human egg is not fundamentally different from that of other animals.

1831 English naturalist Charles Robert Darwin (1809–1882), begins his historic voyage on the H.M.S. *Beagle* (1831–36).

1839 German physiologist Theodore Schwann (1810–1882) states that all living things are made up of cells, each of which contains certain essential components. Schwann's theory is applied to both animals and plants and becomes known as the cell theory.

1858 Modern biology begins as German pathologist Rudolph Virchow (1821–1902) founds cellular pathology with his historic statement that "Every cell comes from a cell."

1859 The landmark book, *On the Origin of Species,* is published by Charles Darwin. This revolutionary work proposes a theory of evolution based on variation and survival of the fittest.

1864 Pasteurization is invented by French chemist Louis Pasteur (1822–1895). Earlier he recognized the relation between microorganisms and disease as well as microorganisms and fermentation.

1866 The laws of inheritance, or genetics, are first stated by Austrian botanist Gregor Johann Mendel (1822–1884). He also states that both male and female contribute equal factors (genes) to the offspring and that these factors do not blend but remain distinct.

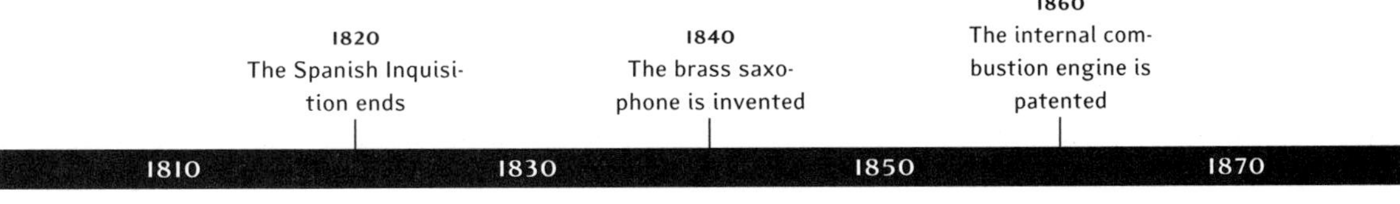

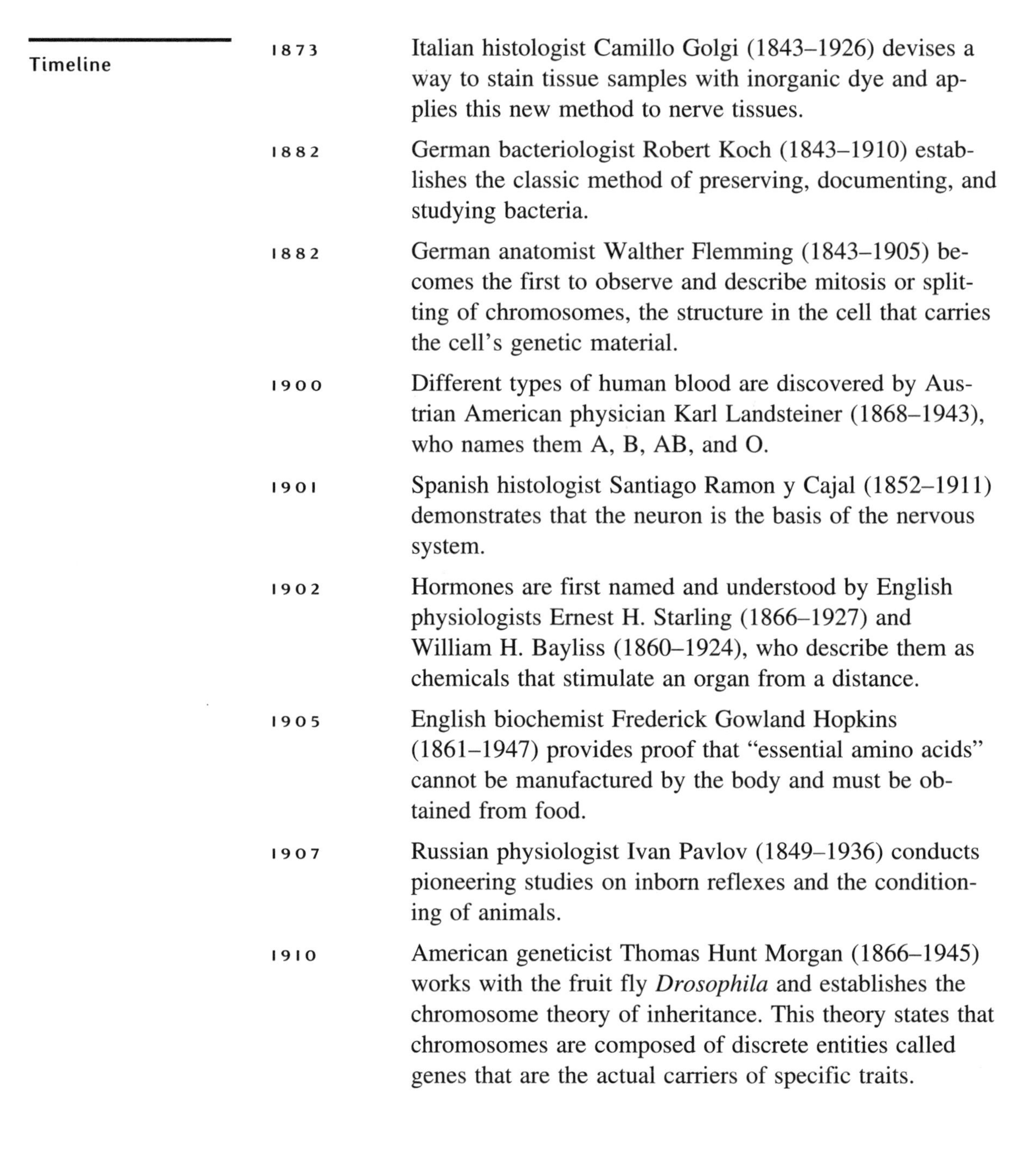

1873 Italian histologist Camillo Golgi (1843–1926) devises a way to stain tissue samples with inorganic dye and applies this new method to nerve tissues.

1882 German bacteriologist Robert Koch (1843–1910) establishes the classic method of preserving, documenting, and studying bacteria.

1882 German anatomist Walther Flemming (1843–1905) becomes the first to observe and describe mitosis or splitting of chromosomes, the structure in the cell that carries the cell's genetic material.

1900 Different types of human blood are discovered by Austrian American physician Karl Landsteiner (1868–1943), who names them A, B, AB, and O.

1901 Spanish histologist Santiago Ramon y Cajal (1852–1911) demonstrates that the neuron is the basis of the nervous system.

1902 Hormones are first named and understood by English physiologists Ernest H. Starling (1866–1927) and William H. Bayliss (1860–1924), who describe them as chemicals that stimulate an organ from a distance.

1905 English biochemist Frederick Gowland Hopkins (1861–1947) provides proof that "essential amino acids" cannot be manufactured by the body and must be obtained from food.

1907 Russian physiologist Ivan Pavlov (1849–1936) conducts pioneering studies on inborn reflexes and the conditioning of animals.

1910 American geneticist Thomas Hunt Morgan (1866–1945) works with the fruit fly *Drosophila* and establishes the chromosome theory of inheritance. This theory states that chromosomes are composed of discrete entities called genes that are the actual carriers of specific traits.

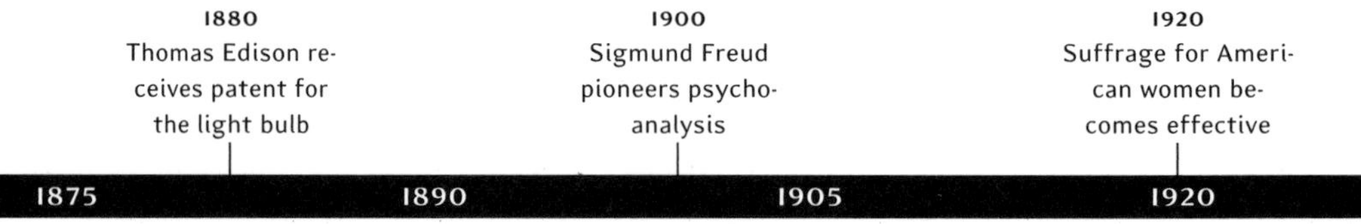

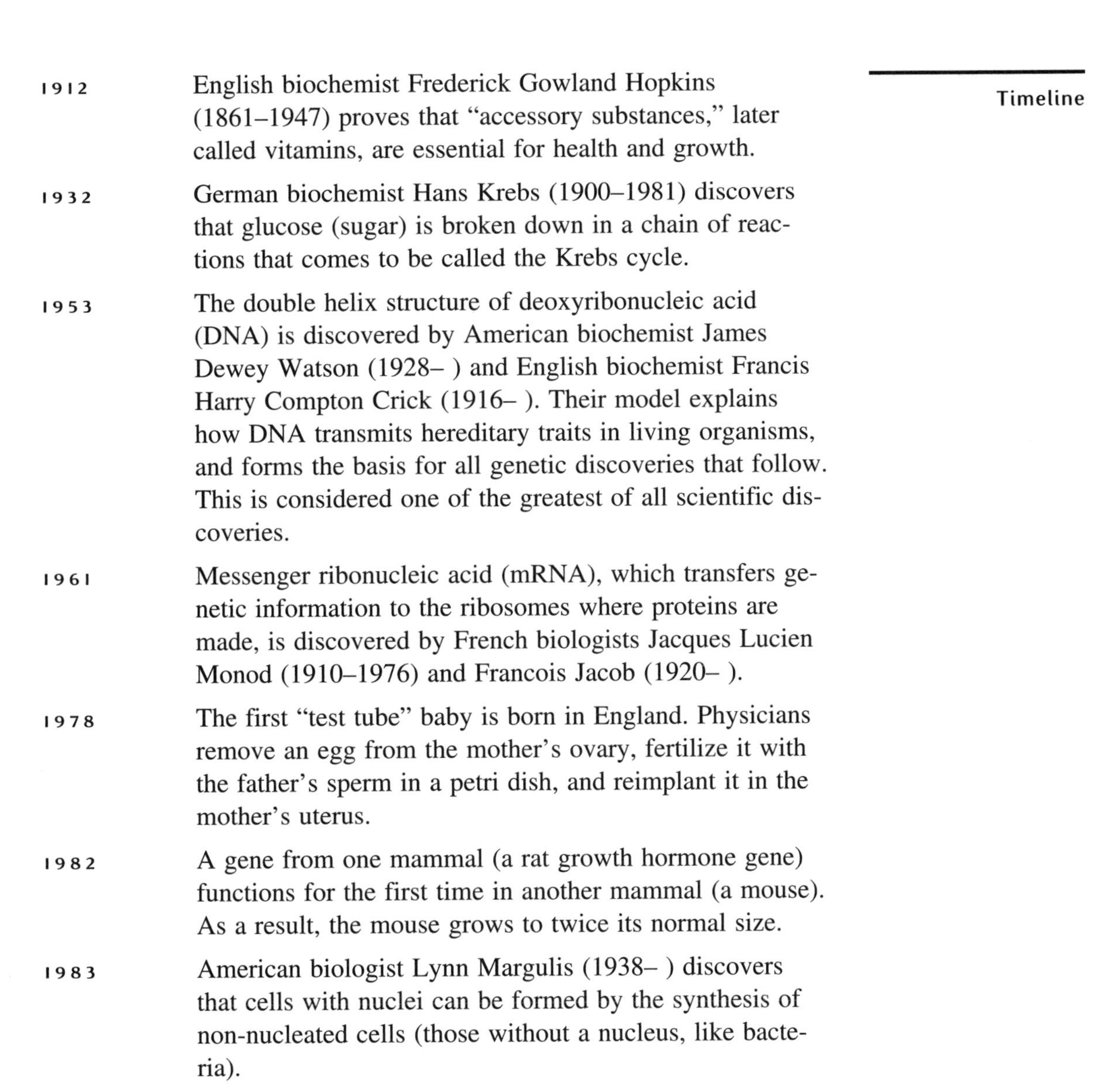

1912 English biochemist Frederick Gowland Hopkins (1861–1947) proves that "accessory substances," later called vitamins, are essential for health and growth.

1932 German biochemist Hans Krebs (1900–1981) discovers that glucose (sugar) is broken down in a chain of reactions that comes to be called the Krebs cycle.

1953 The double helix structure of deoxyribonucleic acid (DNA) is discovered by American biochemist James Dewey Watson (1928–) and English biochemist Francis Harry Compton Crick (1916–). Their model explains how DNA transmits hereditary traits in living organisms, and forms the basis for all genetic discoveries that follow. This is considered one of the greatest of all scientific discoveries.

1961 Messenger ribonucleic acid (mRNA), which transfers genetic information to the ribosomes where proteins are made, is discovered by French biologists Jacques Lucien Monod (1910–1976) and Francois Jacob (1920–).

1978 The first "test tube" baby is born in England. Physicians remove an egg from the mother's ovary, fertilize it with the father's sperm in a petri dish, and reimplant it in the mother's uterus.

1982 A gene from one mammal (a rat growth hormone gene) functions for the first time in another mammal (a mouse). As a result, the mouse grows to twice its normal size.

1983 American biologist Lynn Margulis (1938–) discovers that cells with nuclei can be formed by the synthesis of non-nucleated cells (those without a nucleus, like bacteria).

1987 Genetically engineered plants are first developed.

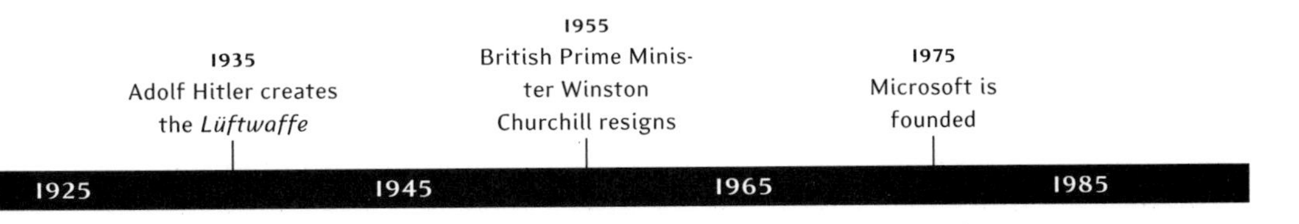

1990 The Human Genome Project is established in Washington, D.C., as an international team of scientists announces a plan to compile a "map" of human genes.

1991 The gender of a mouse is changed at the embryo stage.

1992 The United Nations Conference on Environment and Development is held in Brazil and is attended by delegates from 178 countries, most of whom agree to combat global warming and to preserve biodiversity.

1995 The first complete sequencing of an organism's genetic make up is achieved by the Institute for Genomic Research in the United States. The institute uses an unconventional technique to sequence all 1,800,000 base pairs that make up the chromosome of a certain bacterium.

1997 The first successful cloning of an adult mammal is achieved by Scottish embryologist Ian Wilmut (1944–), who clones a lamb named Dolly from a cell taken from the mammary gland of a sheep.

1998 The first completed genome of an animal, a roundworm, is achieved by a British and American team. The genetic map shows the 97,000,000 genetic letters in correct sequence, taken from the worm's 19,900 genes.

1999 Danish researchers find what they believe is evidence of the oldest life on Earth—fossilized plankton from 3,700,000,000 years ago.

2000 Gene therapy succeeds unequivocally for the first time as doctors in France add working genes to three infants who could not develop their own complete immune systems.

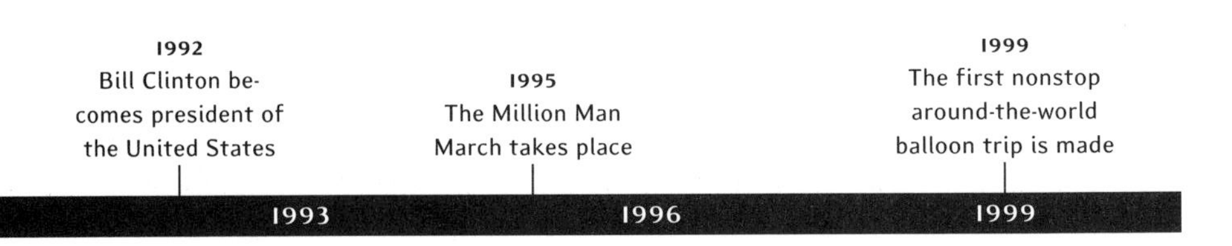

Words to Know

A

Abiotic: The nonliving part of the environment.

Absorption: The process by which dissolved substances pass through a cell's membrane.

Acid: A solution that produces a burning sensation on the skin and has a sour taste.

Acid rain: Rain that has been made strongly acidic by pollutants in the atmosphere.

Acquired characteristics: Traits that are developed by an organism during its lifetime; they cannot be inherited by offspring.

Active transport: In cells, the transfer of a substance across a membrane from a region of low concentration to an area of high concentration; requires the use of energy.

Adaptation: Any change that makes a species or an individual better suited to its environment or way of life.

Adrenalin: A hormone released by the body as a result of fear, anger, or intense emotion that prepares the body for action.

Aerobic respiration: A process that requires oxygen in which food is broken down to release energy.

AIDS: A disease caused by a virus that disables the immune system.

Algae: A group of plantlike organisms that make their own food and live wherever there is water, light, and a supply of minerals.

Allele: An alternate version of the same gene.

Alternation of generations: The life cycle of a plant in which asexual stages alternate with sexual stages.

Amino acids: The building blocks of proteins.

Amoeba: A single-celled organism that has no fixed shape.

Amphibians: A group of vertebrates that spend part of their life on land and part in water; includes frogs, toads, and salamanders.

Anaerobic respiration: A stage in the breaking down of food to release energy that takes place in the absence of oxygen.

Anaphase: The stage during mitosis when chromatids separate and move to the cell poles.

Angiosperms: Flowering plants that produce seeds inside of their fruit.

Anther: The male part of a flower that contains pollen; a saclike container at the tip of the stamen.

Antibiotics: A naturally occurring chemical that kills or inhibits the growth of bacteria.

Antibody: A protein made by the body that locks on, or marks, a particular type of antigen so that it can be destroyed by other cells.

Antigen: Any foreign substance in the body that stimulates the immune system to action.

Arachnid: An invertebrate that has four pairs of jointed walking legs.

Arthropod: An invertebrate that has jointed legs and a segmented body.

Atom: The smallest particle of an element.

Autotroph: An organism, such as a green plant, that can make its own food from inorganic materials.

Auxins: A group of plant hormones that control the plant's growth and development.

Axon: A long, threadlike part of a neuron that conducts nerve impulses away from the cell.

B

Bacteria: A group of one-celled organisms so small they can only be seen with a microscope.

Binomial nomenclature: The system in which organisms are identified by a two-part Latin name; the first name is capitalized and identifies the genus; the second name identifies the species of that genus.

Biological community: A collection of all the different living things found in the same geographic area.

Biological diversity: A broad term that includes all forms of life and the ecological systems in which they live.

Biomass: The total amount of living matter in a given area.

Biome: A large geographical area characterized by distinct climate and soil and particular kinds of plants and animals.

Biosphere: All parts of Earth, extending both below and above its surface, in which organisms can survive.

Biotechnology: The alteration of cells or biological molecules for a specific purpose.

Bipedalism: Walking on two feet; a human characteristic.

Binary fission: A type of asexual reproduction that occurs by splitting into two more or less equal parts; bacteria usually reproduce by splitting in two.

Blood: A complex liquid that circulates throughout an animal's body and keeps the body's cells alive.

Blood type: A certain class or group of blood that has particular properties.

Brain: The control center of an organism's nervous system.

Breeding: The crossing of plants and animals to change the characteristics of an existing variety or to produce a new one.

Bud: A swelling or undeveloped shoot on a plant stem that is protected by scales.

C

Calorie: A unit of measure of the energy that can be obtained from a food; one calorie will raise the temperature of one kilogram of water by one degree Celsius.

Camouflage: Color or shape of an animal that allows it to blend in with its surroundings.

Carbohydrates: A group of naturally occurring compounds that are essential sources of energy for all living things.

Carbon cycle: The process in which carbon atoms are recycled over and over again on Earth.

Carbon dioxide: A major atmospheric gas.

Carbon monoxide: An odorless, tasteless, colorless, and poisonous gas.

Carnivores: A certain family of mammals that have specially shaped teeth and live by hunting.

Carpel: The female organ of a flower that contains its stigma, style, and ovary.

Cartilage: Smooth, flexible connective tissue found in the ear, the nose, and the joints.

Catalyst: A substance that increases the speed at which a chemical reaction occurs.

Cell: The building block of all living things

Cell theory: States that the cell is the basic building block of all life-forms and that all living things, whether plants or animals, consist of one or more cells.

Cellulose: A carbohydrate that plants use to form the walls of their cells.

Central nervous system: The brain and spinal cord of a vertebrate; it interprets messages and makes decisions involving action.

Centriole: A tiny structure found near the nucleus of most animal cells that plays an important role during cell division.

Cerebellum: The part of the brain that coordinates muscular coordination and balance; the second largest part of the human brain.

Cerebrum: The part of the brain that controls thinking, speech, memory, and voluntary actions; the largest part of the human brain.

Cetacean: A mammal that lives entirely in water and breathes air through lungs.

Chlorophyll: The green pigment or coloring matter in plant cells; it works by transferring the Sun's energy in photosynthesis.

Chloroplast: The energy-converting structures found in the cells of plants.

Chromatin: Ropelike fibers containing deoxyribonucleic acid (DNA) and proteins that are found in the cell nucleus and which contract into a chromosome just before cell division.

Chromosome: A coiled structure in the nucleus of a cell that carries the cell's deoxyribonucleic acid (DNA).

Cilia: Short, hairlike projections that can beat or wave back and forth; singular, cilium.

Classification: A method of organizing plants and animals into categories based on their appearance and the natural relationships between them.

Cleavage: Early cell division in an embryo; each cleavage approximately doubles the number of cells.

Cloning: A group of genetically identical cells descended from a single common ancestor.

Cnidarian: A simple invertebrate that lives in the water and has a digestive cavity with only one opening.

Cochlea: A coiled tube filled with fluid in the inner ear whose nerve endings transmit sound vibrations.

Community: All of the populations of different species living in a specific environment.

Conditioned reflex: A type of learned behavior in which the natural stimulus for a reflex act is substituted with a new stimulus.

Consumers: Animals that eat plants who are then eaten by other animals.

Cornea: The transparent front of the eyeball that is curved and partly focuses the light entering the eye.

Cranium: The dome-shaped, bony part of the skull that protects the brain; it consists of eight plates linked together by joints.

Crustacean: An invertebrate with several pairs of jointed legs and two pairs of antennae.

Cytoplasm: The contents of a cell, excluding its nucleus.

D

Daughter cells: The two new, identical cells that form after mitosis when a cell divides.

Decomposer: An organism, like bacteria and fungi, that feed upon dead organic matter and return inorganic materials back to the environment to be used again.

Dendrite: Any branching extension of a neuron that receives incoming signals.

Deoxyribonucleic acid (DNA): The genetic material that carries the code for all living things.

Differentiation: The specialized changes that occur in a cell as an embryo starts to develop.

Diffusion: The movement or spreading out of a substance from an area of high concentration to the area of lowest concentration.

Dominant trait: An inherited trait that masks or hides a recessive trait.

Double helix: The "spiral staircase" shape or structure of the deoxyribonucleic acid (DNA) molecule.

E

Ecosystem: A living community and its nonliving environment.

Ectoderm: In a developing embryo, the outermost layer of cells that eventually become part of the nerves and skin.

Ectotherm: A cold-blooded animal, like a fish or reptile, whose temperature changes with its surroundings.

Element: A pure substance that contains only one type of atom.

Endangered species: Any species of plant or animal that is threatened with extinction.

Endoderm: In a developing embryo, the innermost layer of cells that eventually become the organs and linings of the digestive, respiratory, and urinary systems.

Endoplasmic reticulum: A network of membranes or tubes in a cell through which materials move.

Endotherm: A warm-blooded animal, like a mammal or bird, whose metabolism keeps its body at a constant temperature.

Energy: The ability to do work.

Enzyme: A protein that acts as a catalyst and speeds up chemical reactions in living things.

Epidermis: The outer layer of an animal's skin; also the outer layer of cells on a leaf.

Eukaryote: An organism whose cells contain a well-defined nucleus that is bound by a membrane.

Eutrophication: A natural process that occurs in an aging lake or pond as it gradually builds up its concentration of plant nutrients.

Evolution: A scientific theory stating that species undergo genetic change over time and that all living things originated from simple organisms.

Exoskeleton: A tough exterior or outside skeleton that surrounds an animal's body.

Extinction: The dying out and permanent disappearance of a species.

F

Fermentation: A chemical process that breaks down carbohydrates and other organic materials and produces energy without using oxygen.

Fertilization: The union of male and female sex cells.

Fetus: A developing embryo in the human uterus that is at least two months old.

Flagella: Hairlike projections possessed by some cells that whip from side to side and help the cell move about; singular, flagellum.

Food chain: A sequence of relationships in which the flow of energy passes.

Food web: A network of relationships in which the flow of energy branches out in many directions.

Fossil: The preserved remains of a once-living organism.

Fruit: The mature or ripened ovary that contains a flower's seeds.

Fungi: A group of many-celled organisms that live by absorbing food and are neither plant nor animal.

G

Gaia hypothesis: The idea that Earth is a living organism and can regulate its own environment.

Gamete: Sex cells used in reproduction; the ovum or egg cell is the female gamete and the sperm cell is the male gamete.

Gastric juice: The digestive juice produced by the stomach; it contains weak hydrochloric acid and pepsin (which breaks down proteins).

Gene: The basic unit of heredity.

Genetic code: The information that tells a cell how to interpret the chemical information stored inside deoxyribonucleic acid (DNA).

Genetic disorder: Conditions that have some origin in a person's genetic makeup.

Genetic engineering: The deliberate alteration of a living thing's genetic material to change its characteristics.

Genetic theory: The idea that genes are the basic units in which characteristics are passed from one generation to the next.

Genetic therapy: The process of manipulating genetic material either to treat a disease or to change a physical characteristic.

Genotype: The genetic makeup of a cell or an individual organism; the sum total of all its genes.

Geolotic record: The history of Earth as recorded in the rocks that make up its crust.

Germination: The earliest stages of growth when a seed begins to transform itself into a living plant that has roots, stems, and leaves.

Gland: A group of cells that produce and secrete enzymes, hormones, and other chemicals in the body.

Golgi body: A collection of membranes inside a cell that packages and transports substances made by the cell.

Greenhouse effect: The name given to the trapping of heat in the lower atmosphere and the warming of Earth's surface that results.

Gymnosperm: Plants with seeds that are not protected by any type of covering.

H

Habitat: The distinct, local environment where a particular species lives.

Heart: A muscular pump that transports blood throughout the body.

Hemoglobin: A complex protein molecule in the red blood cells of vertebrates that carries oxygen molecules in the bloodstream.

Herbivore: Animals that eat only plants.

Herpetology: The scientific study of amphibians and reptiles.

Heterotroph: An organism, like an animal, that cannot make its own food and must obtain its nutrients be eating plants or other animals.

Hibernation: A special type of deep sleep that enables an animal to survive the extreme winter cold.

Homeostasis: The maintenance of stable internal conditions in a living thing.

Hominid: A family of primates that includes today's humans and their extinct direct ancestors.

Hormones: Chemical messengers found in both animals and plants.

Host: The organism on or in which a parasite lives.

Hybrid: The offspring of two different species of plant or animal.

Hypothesis: A possible answer to a scientific problem; it must be tested and proved by observation and experiment.

I

Ichthyology: The branch of zoology that deals with fish.

Immunization: A method of helping the body's natural immune system be able to resist a particular disease.

Inbreeding: The mating of organisms that are closely related or which share a common ancestry.

Instincts: A specific inborn behavior pattern that is inherited by all animal species.

Interphase: The stage during mitosis when cell division is complete.

Invertebrates: Any animal that lacks a backbone, such as paramecia, insects, and sea urchins.

Iris: The colored ring surrounding the pupil of the vertebrate eye; its muscles control the size of the pupil (and therefore the amount of light that enters).

K

Karyotype: A diagnostic tool used by physicians to examine the shape, number, and structure of a person's chromosomes when there is a reason to suspect that a chromosomal abnormality may exist.

L

Lactic acid: An organic compound found in the blood and muscles of animals during extreme exercise.

Larva: The name of the stage between hatching and adulthood in the life cycle of some invertebrates.

Lipids: A group of organic compounds that include fats, oils, and waxes.

Lysosome: Small, round bodies containing digestive enzymes that break down large food molecules into smaller ones.

M

Malnutrition: The physical state of overall poor health that can result from a lack of enough food to eat or from eating the wrong foods.

Mammals: A warm-blooded vertebrate with some hair that feeds milk to its young.

Medulla: The part of the brain just above the spinal cord that controls certain involuntary functions like breathing, heartbeat rate, sneezing, and vomiting; the smallest part of the brain.

Meiosis: A specialized form of cell division that takes place only in the reproductive cells.

Membrane: A thin barrier that separates a cell from its surroundings.

Mendelian laws of inheritance: A theory that states that characteristics are not inherited in a random way but instead follow predictable, mathematical patterns.

Mesoderm: In a developing embryo, the middle layer of cells that eventually become bone, muscle, blood, and reproductive organs.

Metabolism: All of the chemical processes that take place in an organism when it obtains and uses energy.

Metamorphosis: The extreme changes that some organisms go through when they pass from an egg to an adult.

Metaphase: The stage during mitosis when the chromosomes line up across the center of the spindle.

Microorganism: Any form of life too small to be seen without a microscope, such as bacteria, protozoans, and many algae; also called microbe.

Migration: The seasonal movement of an animal to a place that offers more favorable living conditions.

Mineral: An inorganic compound that living things need in small amounts, like potassium, sodium, and calcium.

Mitochondria: Specialized structures inside a cell that break down food and release energy.

Mitosis: The division of a cell nucleus to produce two identical cells.

Molars: Chewing teeth that grind or crush food; the back teeth in the jaws of mammals.

Molecule: A chemical unit consisting of two or more linked atoms.

Mollusk: A soft-bodied invertebrate that is often protected by a hard shell.

Molting: The shedding and discarding of the exoskeleton; some insects molt during metamorphosis, and snakes shed their outer skin in order to grow larger.

Monerans: A group of one-celled organisms that do not have a nucleus.

Mutation: A change in a gene that results in a new inherited trait.

N

Natural selection: The process of survival and reproduction of organisms that are best suited to their environment.

Neuron: An individual nerve cell; the basic unit of the nervous system.

Niche: The particular job or function that a living thing plays in the particular place it lives.

Nitrogen cycle: The stages in which the important gas nitrogen is converted and circulated from the nonliving world to the living world and back again.

Nucleic acid: A group of organic compounds that carry genetic information.

Nutrients: Substances a living thing needs to consume that are used for growth and energy; for humans they include fats, sugars, starches, proteins, minerals, and vitamins.

Nutrition: The process by which an organism obtains and uses raw materials from its environment in order to stay alive.

O

Omnivore: An animal that eats both plants and other animals.

Organ: A structural part of a plant or animal that carries out a certain function and is made up of two or more types of tissue.

Organelle: A tiny structure inside a cell that performs a particular function.

Organic compound: Substances that contain carbon.

Organism: Any complete, individual living thing.

Ornithology: The branch of zoology that deals with birds.

Osmosis: The movement of water from one solution to another through a membrane or barrier that separates the solutions.

Oviparous: Term describing an animal that lays or spawns eggs which then develop and hatch outside of the mother's body.

Ovoviparous: Term describing an animal whose young develop inside the mother's body, but who receive nourishment from a yolk and not from the mother.

Oxidation: An energy-releasing chemical reaction that occurs when a substance is combined with oxygen.

Ozone: A form of oxygen found naturally in the stratosphere or upper atmosphere that shields Earth from the Sun's harmful ultraviolet radiation.

P

Paleontology: The scientific study of the animals, plants, and other organisms that lived in prehistoric times.

Parasite: An organism that lives in or on another organism and benefits from the relationship.

Ph: A number used to measure the degree of acidity of a solution.

Phenotype: The outward appearance of an organism; the visible expression of its genotype.

Pheromones: Chemicals released by an animal that have some sort of effect on another animal.

Photosynthesis: The process by which plants use light energy to make food from simple chemicals.

Physiology: The study of how an organism and its body parts work or function normally.

Pistil: The female part of a flower made up of organs called carpels; located in the center of the flower, parts of it become fruit after fertilization.

Plankton: Tiny, free-floating organisms in a body of water.

Pollen: Dustlike grains produced by a flower's anthers that contain the male sex cells.

Pollution: The contamination of the natural environment by harmful substances that are produced by human activity.

Population: All the members of the same species that live together in a particular place.

Predator: An organism that lives by catching, killing, and eating another organism.

Primate: A type of mammal with flexible fingers and toes, forward-pointing eyes, and a well-developed brain.

Producer: A living thing, like a green plant, that makes its own food and forms the beginning of a food chain, since it is eaten by other species.

Prokaryote: An organism, like bacteria or blue-green algae, whose cells lack both a nucleus and any other membrane-bound organelles.

Research and Activity Ideas

Activity 1: Studying an Ecosystem

Ecosystems are everywhere—your backyard, a nearby park, or even a single, rotting log. To study an ecosystem, you need only choose an individual natural community to observe and study and then begin to keep track of all of the interactions that occur among the living and nonliving parts of the ecosystem. Look carefully and study the entire ecosystem, deciding on what its natural boundaries are. Making a map or a drawing on graph paper of the complete site always helps. Next, you should classify the major biotic (living) and abiotic (nonliving) factors in the ecosystem and begin to observe the organisms that live there. Binoculars sometimes help to observe distant objects or to keep from interfering with the activity. A small magnifying glass is also useful for studying small creatures. You should also search for evidence of creatures that you do not see. A camera is also useful sometimes, especially when comparing the seasonal changes in an ecosystem. It is very important to keep a notebook of your observations, keeping track of any creatures you find and where you find them. You can learn more about your ecosystem by counting the different populations discovered there, as well as classifying them according to their ecosystem roles like producer, consumer, or decomposer. A diagram can then be made of the ecosystem's food web. You can search for evidence of competition as well as other types of relationships such as predator-prey or parasitism. You can even keep a record of changes such as plant or animal growth, the birth of offspring, or weather fluctuations. Finally, you can try to predict what might happen if some part of the ecosystem were disturbed or greatly changed. Ecosystems themselves are related to other ecosystems in many ways, and it is important to always realize that all the living and nonliving things on Earth are ultimately connected to one another.

Activity 2: Studying the Greenhouse Effect

The greenhouse effect is the name given to the natural trapping of heat in the lower atmosphere and the warming of Earth's surface that results. This global warming is a natural process that keeps our planet warm and hospitable to life. However, when this normal process is exaggerated or enhanced because of certain human activities, too much heat can be trapped and the increased warming could result in harmful climate changes.

The greenhouse effect can be produced by trying the following experiment. Using two trays filled with moist soil and some easy-to-grow seeds like beans, place a flat thermometer on the soil surface of each tray. After inserting tall wooden skewers in the four corners of one tray, cover it completely with plastic wrap and secure it with a large rubber band. Leave the other tray uncovered and place both trays outside where they are sheltered from the rain but exposed to the Sun. Record the temperature of each tray at the same time each day and note all the differences between the plants. The plastic-wrapped tray should be warmer and its seedling plants should grow larger. This is evidence of the beneficial aspects of the greenhouse effect. However, if the plastic wrap is left over the seedlings for too long they will overheat, wither, and die.

Activity 3: Studying Photosynthesis

If you have ever picked up a piece of wood that has been sitting on the grass for some time and noticed that the patch underneath has lost its greenness and appears yellow or whitish, you have witnessed the opposite of photosynthesis. Since a green plant cannot exist without sunlight, when it is left totally in the dark, the chlorophyll departs from its leaves and photosynthesis no longer takes place. The key role of sunlight can be easily demonstrated by germinating pea seeds and placing them in pots of soil. After placing some pots in a place where they will receive plenty of direct sunlight, place the other pots in a very dark area. After a week to ten days, compare the seedlings in the sunlight to those left in the dark. The root structure of both is especially interesting.

Another way of demonstrating the importance of sunlight to a plant is to pick a shrub, tree, or houseplant that has large individual leaves. Using aluminum foil or pieces of cardboard cut into distinct geometrical shapes that are small enough not to cover the entire leaf but large enough to cover at least half, paperclip each shape to a different leaf. After about a week, remove the shapes from the leaves and compare what you see now to those leaves that were not covered. The importance of sunlight will be dramatically noticeable.

Finally, as a way of demonstrating the exchange of gases (carbon dioxide and oxygen) that occurs during photosynthesis, place a large glass over some potted pea seedlings and place them in sunlight. In time, you will notice that some liquid has condensed on the inside of the glass. This condensation is water vapor that has been given off by the plant when it exchanges oxygen for the carbon dioxide it needs.

Activity 4: Studying Osmosis

In the life sciences, osmosis occurs at the cellular level. For example, in mammals it plays a key role in the kidneys, which filter urine from the blood. Plants also get the water they need through osmosis that occurs in their root hairs. Everyday examples of osmosis can be seen when we sprinkle sugar on a grapefruit cut in half. We notice that the surface becomes moist very quickly and a sweet syrup eventually forms on its top surface. Once the crystallized sugar is dissolved by the grapefruit juices and becomes a liquid, the water molecules will automatically move from where they are greater in number to where they are fewer, so the greater liquid in the grapefruit forms a syrup with the dissolved sugar. Placing a limp stalk of celery in water will restore much of its crispness and gives us another example of osmosis.

Osmosis occurs in plants and animals at the cellular level because their cell membranes are semipermeable (meaning that they will allow only molecules of a certain size or smaller to pass through them). Osmosis can be studied directly by observing how liquid moves through the membrane of an egg. This requires that you get at an egg's membrane by submerging a raw egg (still in its shell) completely inside a wide-mouth jar of vinegar. Record the egg's weight and size (length and diameter) before doing this. The acetic acid in the vinegar will eventually dissolve the shell because the shell is made of calcium carbonate or limestone which reacts with acid to produce carbon dioxide gas. You will observe this gas forming as bubbles on the surface. After about 72 hours, the shell should be dissolved but the egg will remain intact because of its transparent membrane.

After carefully removing the egg from the jar of vinegar, weigh and measure the egg again. You will notice that its proportions have increased. The egg has gotten larger because the water in the vinegar moved through the egg's membrane into the egg itself (because of the higher concentration of water in the vinegar than in the egg). The contents of the egg did not pass out of the membrane since the contents is too large.

The opposite of this activity can be performed using thick corn syrup instead of water. If the egg has its shell removed in the same manner as above but is then immersed for about 72 hours in a jar of syrup, you will find that the egg will have shrunken noticeably. This is because the water concentration of the syrup outside the egg is much less than that inside the egg, so the membrane allows water to move from the egg to the syrup.

Activity 5: Studying Inherited Traits

An inherited trait is a feature or characteristic of an organism that has been passed on to it in its genes. This transmission of the parents' traits to their offspring always follows certain principles or laws. The study of how these inherited traits are passed on is called genetics. Genetics influences everything about us, including the way we look, act, and feel, and some of our inherited traits are very noticeable. Besides these very obvious traits like hair and skin color, there are certain other traits that are less noticeable but very interesting. One of these is foot size. Another is free or attached earlobes. Still another is called "finger hair."

All of these are traits that are passed from parents to their offspring. You can collect data on any particular inherited characteristic and therefore learn more about how genetics works. You will need to collect data about each trait and develop a chart. Any of the above inherited traits can be analyzed. For example, there are generally two types of earlobes. They may be free, and therefore hang down below where the earlobe bottom joins the head, or they may be attached and have no curved bottom that appears to hang down freely. Foot length is simply the size of your own foot and is measured from the tip of the big toe to the back of the heel. The finger hair trait always appears in one of two forms. It is either there or it isn't. People who have the finger hair trait have some hair on the middle section of one or more fingers (which is the finger section between the two bendable joints of your finger).

In order to study one of these interesting traits like finger hair or type of earlobes, you should construct a table or chart that records data on the trait for as many of your family members as you wish. Although it is best to include a large sampling, such as starting with both sets of grandparents and working through any aunts, uncles, and cousins you can contact, even a small sample with only a few members can be helpful. Once you have determined the type of trait each family member has, you should draw your family's "pedigree" for that trait. This is simply a diagram of connected individuals that looks like any other genealogical diagram (which starts at the top with two parents and draws a line from them down to their offspring, and so on). You should use some sort of easily identifiable code or color to signify which individual has or does not have a certain trait. The standard coding technique for tracing the occurrence of a trait in a family is to represent males by squares and females by circles. Usually, a solid circle or square means that a person has the trait, while an empty square or circle shows they do not. In more elaborate pedigrees, a half-colored circle or square means that the person is a carrier but does not show the trait. Once you have done your pedigree, you may do the same for a friend's family and compare his or her family's distribution of the same trait. By comparing the two families' pedigrees for the same trait, you may be able to find certain general patterns of inheritance and to answer certain basic questions. For example, in studying the finger hair trait, you may be able to answer the question whether or not both parents must have finger hair for their offspring to also have it. You might also discover whether both parents having finger hair means that *every* offspring must show the same trait.

U·X·L
COMPLETE
LIFE
SCIENCE
RESOURCE

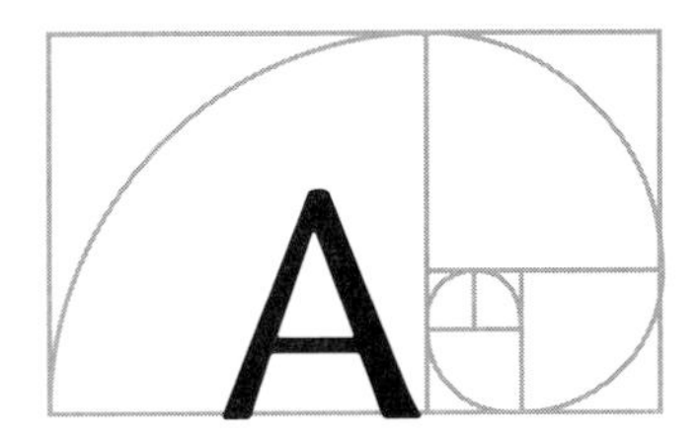

Abiotic/Biotic Environment

The environment or surroundings in which every organism, or living thing, exists can be divided into two categories: its abiotic and its biotic environment. Abiotic refers to the nonliving part of the environment, while biotic refers to its living part.

Typical examples of abiotic factors that make up an organism's environment are sun, wind, rain, soil, and water. Examples of biotic factors are any of the organisms, such as bacteria, plants, and animals that live in the environment. Abiotic forces or factors affect and influence life in an environment. Dramatic examples of abiotic factors are acid rain or a severe storm, while more ordinary examples are water temperature or the amount of oxygen dissolved in a stream. Plants are most affected by abiotic factors including quantity of minerals in the soil, the amount of sunlight received by the plant, and the effects of wind, water, and overall temperature of climate. Animals are less affected by abiotic factors, although temperature and the availability of water can impact an animal's ability to survive.

While an abiotic environment consists of nonliving things, biotic factors include all of the living things, plant and animal, that might be found in any given environment. While green plants get much of the energy they need from the abiotic factors in their environment (sunlight, water, minerals), most other living things (like animals) depend more on biotic elements (since these animals must eat plants and/or other living things to survive).

The organisms that make up the biotic part of an environment can interact in several different ways. Organisms with a neutral interaction have little or no contact with or impact on each other. Organisms may compete for the same resource. They can have a predatory relationship (in which one species benefits by killing another); they can have a parasitic relationship (in which one benefits by living off another); or they can have a mutually beneficial relationship that is favorable to both. Similarly, abiotic factors can influence and shape one another, just as the wind shapes a dune or sunlight heats a pond.

Any environment that contains and supports living organisms is necessarily complex, with living and nonliving elements influencing and modifying each other. Virtually no major environment, or ecosystem, is completely biotic or abiotic, and nearly every living and nonliving element in an environment is related in some way to another. For example, animals and plants (biotic elements) can be affected by climate or weather (abiotic elements); and major changes in the abiotic environment (such as increased rainfall and flooding or a severe and prolonged drought) can alter the conditions and threaten the existence of certain organisms. A drought (abiotic factor) can eliminate many types of green plants, resulting in a decrease in animal population (due to scarcity of food, cover, and shelter). Conversely, too many animals in one area (biotic factor) can destroy important plant life; causing depletion of plant life, erosion, and the possibility of a changed local environment that will not support any life.

In a properly balanced environment, abiotic and biotic factors work together and make up a healthy, functional system. In a typical environment like a small pond, algae, plants, and animals make up the biotic forces, while the pond's water, minerals, and soil (as well as the amount of light it receives) make up some of the more obvious abiotic factors. Altogether, these interacting parts of an environment that are living and nonliving make up what is called an ecological system or ecosystem.

[*See also* **Biological Community; Ecosystem**]

Acid and Base

Acid and base are terms used by chemists to describe the very different actions and opposing properties of certain chemicals when they are dissolved in water. A solution of acid produces a burning sensation on the skin and has a sour taste. A base solution feels slippery and tastes bitter. Both occur naturally and some acids are essential to life.

Acids and bases are both biologically important compounds. As compounds, however, they are complete opposites because of what happens when they are put into a solution, or liquid mixture. At the subatomic level of protons and electrons, any substance that releases a hydrogen ion (a positively charged hydrogen atom) when in solution is called an acid. Conversely, any substance that combines with or gains a hydrogen ion in solution is called a base. While this may sound complicated, many acids and bases are actually quite familiar to us. The vinegar used in a salad dressing or in ketchup and pickles gets its flavor from acetic acid. We have only to place an aspirin on our tongues for a moment to recognize its distinctly acidic, sour taste. We also know immediately that lemons and other citrus fruit contain some form of natural acid. On the other hand, when we wash with soap, we can feel the slipperiness of a base substance, and when we take an antacid tablet for an upset stomach we can experience its neutralizing effect on acids.

People have been using both acids and bases ever since they first started making food and drink. When wine turns sour it changes to vinegar, a diluted or weak form of acetic acid. Spilling some vinegar, or even lemon juice, on a cut lets us feel immediately that it is a mildly burning acid. A common base material that is found in nature is limestone, which people eventually learned to roast and obtain lime. Today we sprinkle this white powder on some soils that are too "acidic" for certain plants to thrive.

Acids come in a variety of strengths, from the fiercely strong hydrochloric acid found in the human stomach, to the mild strength of tomatoes, and the very mild strength found in our own saliva. Strong acids however, are poisonous and can cause severe burns. Acid and base strength are measured on a pH scale ranging from 1 (strongest acid) to 14 (strongest base). Since the strength of a particular substance depends on its concentration of hydrogen ions and whether it releases or attracts them in solution, the pH scale was devised to measure this concentration of hydrogen ions. On a scale of 1 to 14, a solution with a pH of 7 is considered neutral. Very potent acids like sulfuric acid and hydrochloric acid have a pH of 1; lemon juice has a pH of 2; vinegar has a pH of 3; tomatoes are 4; black coffee is 5; and urine is 6. Pure water (not rainwater which is slightly acidic) has a neutral pH of 7. Pure water is between the highest acid (1) and the highest base (14), and is actually neither.

Continuing on the pH scale to the base side, seawater has a pH of 8; baking soda has a pH of 9; milk of magnesia is 10; household ammonia is 11; lime is 12; hair remover is 13; and lye is 14. On this scale, a change of only 1 point means a change of ten times the concentration. Thus, lemon juice is ten times more acidic than vinegar, and stomach acid is ten times more acidic than lemon juice. An easy way of distinguishing acids from bases is to use litmus paper. It was discovered that certain organic extracts, such as litmus—which is obtained from lichen, a plant-fungus organism—turns red when dipped in acid and blue when dipped in a base.

Acids play a major biological role at the most fundamental level, since the organic acids called amino acids are necessary for life. Amino acids are known to be the building blocks of proteins. The pH level is also important to life, since every cell is sensitive to it and will not tolerate too great a change from its proper pH level. Because of this, most living systems have several mechanisms to make sure that their internal pH remains fairly constant. An organism's habitat or external environment must also have a range of pH that is suited to it or it will suffer or even die. Acid rain, which results from air pollution, has damaged certain vulnerable forests. It has also made some lakes too acidic (as much as a pH of 5) for trout and other fish. Finally, both acids and bases have been put to a number of practical uses in industry and in the production of consumer goods. We have acid in our batteries and bases in our soaps.

Acid Rain

Acid rain is rain that has been made strongly acidic by pollutants in the atmosphere. It is caused by the burning of fossil fuels and results in declining fish populations and damaged and dead trees. It also damages buildings and statues and aggravates respiratory conditions.

Precipitation or rainwater has always been slightly acidic since water naturally dissolves atmospheric carbon dioxide, but the precipitation described as acid rain often shows an alarmingly high concentration of dangerous acids like sulfuric acid and nitric acid. These powerful acids are in the atmosphere because of human activities, despite the fact that no one deliberately or even accidentally put them there. Rather, acid rain is the indirect result of human technological and industrial progress. Automobiles, factories, and power plants usually burn fossil fuels such as oil, gas, natural gas, or coal. Ever since the Industrial Revolution began in the eighteenth century, scientists have noticed a connection between air pollution, acid rain, and downwind damage to animals and plants. The term acid rain was first used in England around 1870.

HOW ACID RAIN IS FORMED

Acid rain is formed in Earth's atmosphere by a very simple and natural chemical process. First the atmosphere receives a steady dose of gases, such as sulfur and nitrogen oxides, that result from burning huge quantities of fossil fuels. Gas-burning cars and coal-burning electric power plants are two examples of items that give off sulfur dioxide and nitrogen oxides. Sulfur dioxide is an especially smelly toxic gas given off when coal is burned. These gases rise into the sky where they usually react with water, oxygen, and sunlight. It is mainly the moisture in the clouds, after reacting with these gases, that creates nitric acid and sulfuric acid. Sulfuric acid is an especially strong acid and is responsible for well over half of the extra acidity found in acid rain. Nitric acid makes up most of the rest of acid rain, and it too is a very caustic acid.

HOW ACIDIC IS ACID RAIN?

Normal rain has a slightly acidic pH of 5.6. This is on a typical pH acid/base scale of 0 to 14 where 7 indicates neutrality. Numbers less than 7 indicate increasing acidity and numbers greater than 7 represent increasing alkalinity. Acid rain has been found to have a pH ranging between 3.4 and 4.5, which is very acidic. This acid-bearing moisture reaches Earth's surface when it rains, snows, or when fog covers the ground. It is known to be particularly harmful to organisms that live in water. Acid rain can also fall to Earth as particles or gases in a process called dry deposition.

EFFECTS OF ACID RAIN

Some regions are more sensitive to acid rain than others, and an area that has highly alkaline soil tends to neutralize much of the acidity, making it less harmful. However, areas that have thin soil on top of granite do not have the same advantages. What happens to the aquatic organisms in these areas is that the ponds, lakes, and streams they live in become acidic to the point where water conditions actually start to kill the inhabitants. The young and small fish die first and are followed by the amphibians (frogs and salamanders). This alters the ecological balance and usually results in a boom in the mosquito population since their natural predators are gone. If the situation continues, the lake will eventually die from a lack of oxygen after becoming full of dead organisms.

For plants, too much acid in the soil eventually reduces soil fertility. More important however, is the effect that acid soil has on a plant's leaves. Since the acid destroys a leaf's waxy outer coating, it takes away both its protection and its ability to photosynthesize (the process by which the plant makes its own food using sunlight). It is easy to notice evergreen forests damaged by acid rain since the trees usually die from the top down. High altitude trees suffer the worst effects since they are exposed to the highest concentration of acid rain. Plants and trees not killed directly by acid rain are put under stress, making them susceptible to other diseases and enemies.

Besides acidifying water bodies and damaging forests, acid rain is linked to corrosion of statues, monuments, and some buildings. Such well-known monuments like the Lincoln Memorial and the Roman Colosseum were built out of limestone and marble and have shown signs of corroding. Both stone types are composed of calcium carbonate which reacts with the acids and slowly "melts" away the surface detail.

Today, acid rain is recognized as an international problem since airborne pollutants do not stop at national boundaries. For example, much of the acid rain that harms the forests of Norway, Sweden, and Finland is blown there by prevailing winds from western and eastern Europe where the rain originated. Beginning with the passage of the 1970 Clean Air Act in the United States, national legislation and international agreements have been passed to reduce the amount of acid rain that is produced in developed countries. New technologies like "scrubbers," whose purpose it is to control the release of sulfur, are now being used in factories and power plants. Automotive emissions also have been controlled somewhat by the use of catalytic converters. Corrective actions like these have improved the situation in certain parts of the world, yet not all nations are making their best efforts, especially when pollution reduction is so expensive. However, it is important that all countries attempt to control their pollution since the future of the world's ecology depends on it.

Spruce and fir trees in the Great Smoky Mountains killed by acid rain. (Reproduced by permission of JLM Visuals.)

[*See also* **Acid and base; Forests; Pollution**]

Adaptation

Adaptation refers to any change that makes a species or an individual better suited to its environment or way of life. All living creatures must be able to adapt to changes in their environment if they are to survive and reproduce. The process of adaptation does not always result in an obvious physical change, but may affect an individual's behavior or even its internal processes.

Adaptation could be described as the theory of evolution in action. This theory was first offered in its best and most complete form in 1859 by the English naturalist, Charles Darwin (1809–1882). Darwin's theory of evolution suggested that all living things are subject to a gradual process of change over a long period of time. Evolution is therefore the process that results in living things changing through successive generations. Darwin also described a mechanism called "natural selection," which is the means through which these hereditary changes are passed on from one generation to another.

Darwin's theory of evolution by natural selection—which is highly important to understanding the life sciences—states that individual organisms possessing certain traits or characteristics that are most suited to a particular environment have a better chance of surviving and therefore of passing these traits on to their offspring. In other words, organisms in possession of favorable traits allows these traits to be "selected" by nature (natural selection) so that these organisms survive and produce young that have the same favorable traits.

Darwin's years of travel, study, and thought led him to make three important observations or conclusions. First, he stated that all living things vary, or that each individual varies slightly from the others of its species. This could be seen in any given species in which differences could be found among any group of the same organisms (such as a flower with an extra petal or a deer with larger-than-average antlers). Second, Darwin suggested that individuals were able to pass certain characteristics on to their young, who inherited them. Modern genetics has proven this to be true. Third, Darwin noted that all life is involved in a struggle for survival. To Darwin, these three observations explained why nature allowed most organisms to produce far more offspring than could ever survive. Offspring that did survive and were able to reproduce, according to Darwin, were usually the ones who possessed certain traits better suited to their environment. These individuals had a better chance of surviving and of producing more individuals like themselves. Over many generations, it was simply nature's "selection" of the individuals with the fittest (or best adapted) characteristics that explained Darwin's theory of evolution. This accounts for the well-known phrase, "survival of the fittest."

"Survival of the fittest" could also be described as "survival of the best adapted." Among any group of individuals or organisms of the same species, there will always be variations or differences (in color, shape, behavior, and even chemical makeup). An adaptation then, is considered to be any variation that makes an organism better suited to its environment. Camouflage is one way nature has of providing protection for an individual, and an organism whose color or shape allows it to blend into its environment is more likely to survive and reproduce than one whose coloration makes it easier to be noticed. Woodpeckers are highly specialized birds and are a good example of the process of adaptation. Their main job is to find and eat insects that live in and beneath the bark of trees. Consequently, those woodpeckers with the most powerful and chisel-like beaks, strong neck muscles for hammering, sturdy skulls, grasping feet, stiff-supporting tail feathers, and long tongues proved best able to survive and pass on these traits.

Darwin first described his theory of evolution by natural selection in 1859 in his classic book, *On the Origin of Species by Means of Natural Selection.* The idea of adaptation is key to Darwin's theory. Darwin distinguishes between genotypic adaptation and phenotypic adaptation. The type of adaptation discussed above—in which an individual possesses a "favorable" gene for a certain characteristic which it passes on to its offspring—is called genotypic or evolutionary adaptation. This type of adaptation is genetic, permanent, and is very different from phenotypic adaptations. In contrast to genotypic adaptations, phenotypic adaptations are traits that are developed during an individual's lifetime. An example of phenotypic, or nongenetic, adaptation might be a certain type of behavior that is learned or developed by an individual. The macaque monkeys in Japan, for example, have learned to wash their food in water, and newborns soon copy this behavior. Although it is not known exactly why the first macaque washed its food, this behavior in not instinctive with them. Rather it is a case of learned behavior.

[*See also* **Evolution; Evolution, Evidence of; Evolutionary Theory**]

Aerobic/Anaerobic

Aerobic and anaerobic are terms used to describe the presence or absence of oxygen. (Anaerobic means "without air.") All living things require energy, and when oxygen is used to metabolize (convert or break down)

HANS ADOLF KREBS

German-British biochemist Hans Krebs (1900–1981) was the first to explain how cells release energy during respiration (the chemical process by which food is broken down to release energy). His discovery of a very complicated chain of reactions, which came to be called the Krebs cycle, explained how cells break down glucose (a common sugar) and obtain needed energy. For this discovery he received the 1953 Nobel Prize in Physiology and Medicine and was knighted by Queen Elizabeth in 1958.

Hans Krebs was born in Hildesheim, Germany and received his medical degree from the University of Hamburg in 1925. His father was a doctor, and Krebs took up his father's specialty as an ear, nose, and throat specialist. It was not long, however, before he realized that he preferred doing research to working with patients, and in 1926 he became an assistant to the noted biochemist, Otto Heinrich Warburg (1883–1970) in Berlin. Warburg studied respiration and would himself win a Nobel Prize in 1931 for his work on that subject. By 1932, Krebs was making a name for himself with his own work on amino acids (the building blocks of protein), but in 1933 the German dictator, Adolf Hitler (1889–1945), was appointed chancellor. This political change in Germany meant that all people of Jewish origin, including Krebs, would be persecuted and eventually sent to concentration camps to be worked to death. That year however, Krebs was able to leave Germany and move to England. There he had the good fortune to work with another Nobel Prize winner, the English biochemist, Frederick Gowland Hopkins (1861–1947), at Sheffield University. It was there that Krebs would discover the process for which he is best known, the citric acid cycle, which is also called the Krebs cycle.

The Krebs cycle is an important step in the process used by cells when they convert food, such as carbohydrates and fats, into usable energy. During this energy-producing process called respiration, one molecule of glucose combines with six molecules of oxygen to produce six molecules of carbon dioxide, six molecules of water, and a considerable amount of energy. Krebs discovered that this does not happen all at once, but that a complicated chain of reactions occurs during which a little of the original energy is released each time. His work revealed that this series of reactions was actually a chain, or a cycle, of events. His explanation of this highly complex cycle proved to be a major breakthrough in biochemistry and in understanding how an animal's metabolism really works. Metabolism is all of the chemical processes (all the building up and breaking down) that takes place in an organism to stay alive and grow. The Krebs cycle focuses specifically on the breaking-down aspects of an animal's metabolism by which energy is released. It explains how, through a series of six chemical reactions that take place inside an animal cell in a recurring loop, food is combined with oxygen to produce the energy needed for life.

organic materials to release energy, it is described as being an aerobic process. When no oxygen is needed to metabolize materials, the process is referred to as anaerobic. The terms aerobic and anaerobic are used mostly to describe types of respiration and types of bacteria.

AEROBIC RESPIRATION

All living cells need a constant supply of energy to power the chemical activities they conduct to support life. The living cells in both plants and animals use glucose, the most common form of sugar, as their energy source or fuel. However, glucose must be broken down during a process called respiration before it will release usable energy. Finally, energy is produced by the cell in one of two ways: aerobic respiration or anaerobic respiration. Aerobic cellular respiration is a process that necessarily involves the use of oxygen. During this process, glucose and oxygen are chemically combined, which is known as oxidation, in the cell's mitochondria (part of the cell that produces energy) to yield energy and to release carbon dioxide and water as waste. During aerobic respiration, one molecule of glucose is combined with six molecules of oxygen to produce six molecules of carbon dioxide and six molecules of water. Aerobic respiration also releases a large amount of energy in the form of energy-carrying molecules called adenosine diphosphate (ADP) and adenosine triphosphate (ATP). The series of events or reactions that occur during aerobic respiration are known as the Krebs cycle. This cycle is named after the German biochemist Hans Krebs (1900–1981), who discovered that glucose is broken down in a chain of reactions. Aerobic respiration, therefore, results in the release of a large amount of energy, but only if oxygen is present. It is in this way in which animals and plants obtain energy.

ANAEROBIC RESPIRATION

Anaerobic respiration is the opposite of aerobic, since it involves a type of respiration that does not involve oxygen. Also called glycolysis (which literally means the splitting of carbohydrates), this process takes place in the cell and is very slow compared the to oxygen-rich process of aerobic respiration. Anaerobic respiration does not result in the production of a great deal of energy since glucose is only partly broken down. Instead, most of the glucose forms new organic compounds such as acid, methane gas, and alcohol. The most common anaerobic reactions take place during the process known as alcoholic fermentation. During this process, microorganisms like bacteria, molds, or yeast change sugar into carbon dioxide and alcohol. In the best known examples of making bread, beer and wine, yeast is used to bring about this conversion without the use of oxygen. Although the anaerobic process produces little immediate energy, its by-products like methane and alcohol contain a significant amount of potential energy and can be used as fuels.

AEROBIC AND ANAEROBIC BACTERIA

Bacteria can also be described as being aerobic or anaerobic. These single-celled microorganisms are among the most abundant living things on Earth and live and feed in many different ways. Bacteria that need oxygen in order to grow are called aerobic bacteria. An example are the bacteria that cause the lung disease tuberculosis. The moist, warm lungs have a steady supply of oxygen and provide an ideal breeding ground for this potentially fatal bacterium. However, many bacteria can only grow in the absence of oxygen and are called anaerobic bacteria. The bacteria that live in soil and those that inhabit the intestinal tracts of mammals carry out anaerobic respiration.

ANAEROBIC RESPIRATION IN THE MUSCLES

Another type of anaerobic respiration occurs in the stressed muscles of an animal when is uses oxygen faster than the blood can supply it. The exhausted muscle quickly switches from aerobic respiration to anaerobic respiration and begins to break down glucose without oxygen. This type of anaerobic respiration results in little energy and produces muscles that ache and eventually shut down. Anaerobic respiration in the muscles produces an acid by-product known as lactic acid. It is the buildup of lactic acid that makes an athlete's muscles "burn" and then stop working, which is why some runners collapse and can barely use their muscles immediately after a race.

[*See also* **Bacteria; Respiration**]

German-British biochemist Hans Krebs discovered the reactions of aerobic respiration, which was later named the Krebs cycle. (Courtesy of The Library Congress.)

Aging

Aging is the gradual loss of function in both cells and the overall organism. The natural process of aging, or senescence, results in bodily changes that make an organism less efficient and eventually contribute to its death. Aging is almost certainly affected by genes, and members of the same species have similar life expectancies.

After living for three weeks as a larva, a mayfly may spend only one day as an adult before it dies. A bird may live up to four years, a frog sometimes up to twenty, a human being can reach one hundred, and a Sierra redwood tree can live as many as four thousand years. Obviously, different species have radically different life spans. (Life span is the maximum time that an individual may live under ideal circumstances.) It is different from life expectancy, which is calculated from the average years lived by individuals of a certain generation. However, it appears that no matter how long or short the life span of an individual organism may be, it generally undergoes a process of getting older that is marked by gradual deterioration of its systems and abilities. In humans, this process becomes very obvious in what is called middle age. By forty or fifty, a person's body begins to act and appear different. The skin becomes less elastic or smooth and permanent wrinkles appear. These people lose muscle tissue and bone hardness, and their vision and hearing gets less sharp. Even their taste buds start to deteriorate.

Eventually all organisms die, and aging can be considered the process through which animals and plants go on their way through their individual life span. However, science has not yet been able to explain exactly why aging occurs. Gerontology, which is the study of all aspects of aging, has no single theory on how or why people age. One theory says that an individual's life span is programmed by his or her genetic inheritance. Some call this the "time-bomb" theory, claiming that each of us has our own genetic clock or clocks that slow down and eventually cause certain cells to die out. The other major theory of aging is that of wear-and-tear. This argument says that cells eventually break down under the constant assaults of heavy use and environmental insults like chemicals and radiation.

EFFECTS OF AGING

Although scientists are unsure of the exact cause or causes, they know very well the effects of aging on the human body. These include wrinkly skin, muscle loss, bone thinness, a less efficient heart, weakened lungs, poorer vision and hearing, decrease in mental quickness, reduced kidney function, and diminished resistance to infection, among many others. These effects appear to be the result of our cells becoming less efficient in their jobs. As we age, our cells do not do as good a job in functions like removing wastes, destroying poisons, repairing genes, and making proteins. As the cells get weaker and weaker, they do their jobs less well, which means that the entire body becomes less and less efficient or healthy. Although old age can have its share of diseases, such as hardening of the arteries, stroke, cancer, and the brain condition known as Alzheimer's disease, it is important to realize that these diseases are not a natural result of the aging process.

Aging occurs in plants as well as animals, and is usually connected to plant growth cycles. Plants that have what is called determinate growth have a built-in time when they stop growing, after which they slowly breakdown and die. Plants that we call annuals and biennials have a programmed time during which they grow, reach a certain size or age, and then wither and die. Plants that continue for a longer period have indeterminate growth. Some perennials can live for years, despite the fact that their above-ground systems die every winter. The below-ground plant stays alive and recovers in the spring. Others, like the common juniper tree, can live for two thousand years.

Although science has yet to pinpoint the exact reason that aging occurs in any organism, it is safe to say that genetics probably plays the largest role in determining the life span of an individual. The next largest role is probably that of our external environment. A toxic environment no doubt puts an enormous strain on all body systems, which inevitably deteriorate. While good living habits like a balanced diet and regular exercise can minimize some of the effects of the aging process, the reality of growing old and less efficient is, so far, an inevitable fact of life.

Agriculture

Agriculture is the art and science of cultivating the soil, growing and harvesting crops, and raising livestock (animals) for human use. As the world's oldest and most important industry, agriculture provides the basic substances necessary to sustain human life. Without agriculture, the development of civilization could have never occurred.

Long before humans knew how to grow their own supply of food, people practiced hunting, fishing, and the simple gathering of edible plants that happened to be growing locally. This type of existence did not encourage people to stay in one place, since they were often forced to move somewhere else when their supply of game diminished or local plants like fruits, nuts, and roots had a poor growing season. As a result of this hunting and gathering, people were more or less nomads who were often forced to leave and search for food. A constant preoccupation with where the next meal was coming from left people little time to develop any other skills besides those related to finding food. Early humans eventually learned how to domesticate, or tame, dogs and used them to help herd the groups of sheep and goats they had captured (as a supply of fresh food). However, as long as humans had to hunt and gather most of their food, they existed on a day-to-day basis, meaning that they could not meaningfully change their lives in any way. The development of agriculture would eventually change nearly every aspect of human society.

The word agriculture comes from two Latin words, *ager* meaning "field" and *cultura* meaning "cultivation." Sometime around 11,000 B.C., certain tribes discovered that plants could be grown from seeds, and the first crops were probably raised from the seeds of grasses that could be used for food, like wild barley and wild wheat. This was the beginning of farming, and these first farmers usually settled where the soil was fertile and easy to till and where there was water close by. These conditions were usually found in river valleys. In many ways, agriculture is ultimately responsible for both group living (in villages or cities), and for the actual location of the cities themselves. Since farming seemed to arise in different parts of the world at about the same time, many scientists believe that a favorable change in Earth's climate may have been responsible for the development of agriculture.

EFFECTS OF AGRICULTURE

Whatever the reason, growing crops and raising livestock would unavoidably change the way human beings lived. First, it provided them with a steady supply of food that increased their chances of survival. Second, it allowed them to establish permanent settlements since they not only had a close and reliable source of food, but they had to constantly take care of and guard it. Also, since far fewer people were required to steadily search for and gather food, people were free to develop arts and skills like pottery, weaving, and leatherwork. Their technology also improved as they developed new agricultural tools like plows pulled by oxen. Altogether, a steady food supply based on farming ultimately led to the development of human civilization, including culture, art, laws, customs, religion, and government. It is not an overstatement to say that agriculture is the foundation upon which all of human society is based.

A single example of how agriculture influenced science and technology is the annual flooding of the Nile River in Egypt. Every year the Nile would overflow its banks and lay down a rich new layer of soil for that year's crops. This was beneficial but it also erased everyone's boundaries. Since it was important to know when the flood would occur as well as whose land was whose, the Egyptians applied their knowledge of astronomy (the science that deals with the study of the Sun, Moon, stars and other celestial bodies) and invented an accurate calendar based on the phases of the moon. They also made advances in surveying (the measurement and description of a region, part, or feature on the Earth's surface) and mathematics, discovering the practical principle behind the Pythagorean theorem (that in any right triangle, the square of the hypotenuse of the triangle is equal to the sum of the squares of the other two sides) long before the Greek mathematician, Pythagoras of Samos (c.580–c.500 B.C.), ever lived.

The civilizations of ancient Egypt, Greece, and Rome were all based on and supported by a base of agriculture. The Romans were a civilization of wealthy city-dwellers whose specialized agriculture allowed them to sell and trade. They used many sound agricultural techniques, like resting the land (not planting for a season) and plowing under crops to enrich the soil. They also practiced selective breeding techniques and produced the first specialized breed of dairy cattle. The Middle Ages (500–1450) was a time of minor agricultural improvements, but it was marked by terrible happenings like the Black Death (an extremely deadly form of bubonic plague that was widespread throughout Europe and Asia in the fourteenth century) that killed millions of people and created chaos on farms as well as in cities. However, the voyages of discovery that

This corn field is the result of advances in agricultural technology that led to the mass production of crops. (Reproduced by permission of Photo Researchers, Inc.)

marked the beginning of the 1400s contributed greatly to a new agricultural diversity as New World (North and South America) crops and methods were brought back to Europe. Many Native American crops were eventually accepted and grown by Europeans, including corn, peanuts, peppers, tomatoes, potatoes, and squash. The Europeans also brought their own seeds, farming methods, and tools to the New World.

It was not until after 1700 that the practice of agriculture began to benefit from a series of scientific discoveries and technological inventions, setting the stage for what has been called the agricultural revolution of the 1700s. One of the most important agricultural changes was the adoption of the new four-crop rotation in England. This new method led to greater food supplies since farmers replaced the old system of three crops a year for three years (with no planting the fourth) with the new system. They began to grow wheat, barley, turnips, and clover in succession without ever resting the soil. This was possible because the clover put nitrogen back into the soil and also provided grazing for cattle and sheep, which fertilized the soil with their waste. Farmers also began to suspect that their livestock could be improved by repeatedly breeding animals with desirable traits.

By the nineteenth century, agricultural tools had improved greatly. The discovery was also made that chemicals known as phosphates were needed by growing plants, and artificial fertilizers were developed by treating rocks that were rich in phosphate with sulfuric acid. By the twentieth century, science and technology had not only provided farmers new power sources for their tools, but gave farmers ways to further improve their livestock while also increasing their harvest using new agricultural chemicals. These modern chemicals include fertilizers (phosphorous, potassium, and nitrogen), insecticides like DDT dichlorodiphenyltrichloroethane, and herbicides for weed control, as well as other chemicals to fight plant diseases. By the beginning of the twenty-first century, however, both science and farmers have learned that these chemicals can have disastrous environmental side effects, as seen in the DDT experience of the 1950s. This insecticide proved toxic to birds that had eaten insects contaminated with the chemical. As a result, the birds were unable to reproduce, and the U.S. government eventually banned most uses of DDT.

Today, although almost half of the world's labor force is employed in agriculture, in highly developed nations like the United States less than four percent of the population is actively engaged in agriculture. Although the human population is steadily increasing and farmland is shrinking in America, the United States produces an agricultural surplus almost every year. This is not the case in all countries, however. For example, some countries in Africa and Central America are not able to produced enough food for their growing populations. As a result, thousands of people in these countries die every year because of lack of food.

AIDS

AIDS, or Acquired Immune Deficiency Syndrome, is a disease caused by a virus that disables the immune system. The virus enters the body through the bloodstream, duplicates itself rapidly, and eventually destroys the body's immune system. This leaves the victim susceptible to other infectious diseases that usually prove fatal.

AIDS is caused by the Human Immunodeficiency Virus (HIV) that was first isolated in 1983. Before 1981, AIDS was unknown, and many health professionals believed that infectious diseases were a thing of the past in developed countries. However, following the discovery in late 1980 of several young homosexual men who had developed rare forms of pneumonia and cancer, health officials in the United States realized they had discovered a new infectious disease that enters the bloodstream and destroys the immune system. From these beginnings, the worldwide estimates of the number of people infected with AIDS has reached 30,000,000. More than 12,000,000 people have already died from AIDS since the beginnings of this worldwide epidemic, and although the number of infected individuals has stabilized in developed countries, the number of AIDS cases has exploded in many African countries.

HOW AIDS SPREADS

AIDS is obviously a contagious disease for which there is no cure. It is spread by a virus and transmitted by entering the bloodstream. This means that AIDS cannot be spread by the type of casual contact that usually takes place between family members and friends. HIV must somehow enter the bloodstream to infect a person, and the most common way for this to happen is through some form of intimate sexual contact that allows bodily fluids from one person to enter that of another. This is what occurs during any type of sexual intercourse or sexual penetration of a person's body. Another way is for an intravenous drug user to share a needle with another person. HIV has also been transmitted to an unborn child by its infected mother, and until programs for blood screening were created, HIV had also been transmitted by blood transfusions.

ROBERT CHARLES GALLO

American virologist (a person specializing in the study of viruses) Robert Gallo (1937–) is credited as the codiscoverer of the human immunodeficiency virus (HIV). Gallo also established that HIV causes acquired immunodeficiency syndrome, or AIDS (a disease caused by a virus that disables the immune system), and developed the blood test for detecting HIV, which is still the main tool in diagnosing the disease. Moreover, Gallo's blood test for HIV has made the blood supply safe.

Robert Gallo was born in Waterbury, Connecticut, and grew up in the same house that his immigrant grandparents bought after they came to the United States from Italy. His father had a welding business, but the dominant theme of young Gallo's family life was the illness and death of his only sibling, his sister Judy. His sister died of childhood leukemia, and this disease brought Gallo into regular contact with the nonfamily member who most influenced his life, Doctor Marcus Cox. During his senior year in high school, an injury forced Gallo off the basketball team and got him thinking about his future. He began to spend time with Cox, and by the time Gallo was ready for college, he knew he wanted a career in biomedical research. Gallo majored in biology at Providence College in Rhode Island, and went on to earn a medical degree from Jefferson Medical College in Philadelphia in 1963.

In 1965 Gallo joined the National Institutes of Health (NIH) to do cancer research, and in 1971 he was appointed head of a new Tumor Cell Biology lab-

AIDS TAKES OVER THE IMMUNE SYSTEM

AIDS is an especially difficult disease because, unlike other forms of infections, it attacks the very system that we use to defend ourselves against outside invasions. The HIV virus is a ribonucleic acid (RNA) virus like many other viruses, such as the flu, the common cold, and measles. However, HIV is also a retrovirus, which makes it quite different and deadly. Viruses contain their own forms of deoxyribonucleic acid (DNA) and RNA (their own genetic code), but a retrovirus contains a special enzyme that enables it to put its own DNA into the DNA of the cell it invades. A retrovirus can then use the infected cell's machinery to continuously reproduce itself and make more and more copies of the retrovirus.

Once inside the body and reproducing, HIV goes to work by attacking the very type of cells that the body automatically uses to fight invaders like viruses. It mainly attacks what are called the T-4 white blood cells and eventually causes their number to dramatically decrease. As these

oratory at NIH. There he studied retroviruses, which are types of viruses that possess the ability to penetrate other cells and to splice their own genetic material into the genes of their hosts, eventually taking over all of the invaded cell's reproductive functions. This research led him to discover the first human cancer virus.

When AIDS began to be recognized in the United States in early 1981 as a new and terrible disease, it was Gallo's pioneering work in the field of human retrovirology that led him to be one of the first scientists to hypothesize that the disease was caused by a virus. In 1982, the National Cancer Institute formed an AIDS task force with Gallo as its head. By 1984, Gallo's team was able to establish that the Human Immunodeficiency Virus caused AIDS, and it developed a blood test for detecting the virus. It was then that a controversy developed which would involve Gallo for the next decade. The controversy began when a French colleague who had earlier and independently discovered the same virus sent Gallo a sample. The result involved legal disputes and hearings, accusations, findings, and reversals of opinions on which scientists had actually discovered the virus first. Overall, it is now thought that the virus sent to Gallo from France may have contaminated the blood samples that he held, making for a mistake but not misconduct. Today, Gallo has survived and overcome the allegations and is the director of the Institute of Human Virology in Baltimore, Maryland. There, Gallo continues his pioneering work in the field of human retroviruses. Gallo has received many honors, including the distinction of being the most referenced scientist in the world between 1980 and 1990.

important disease-fighting cells are killed off, the body's ability to resist infection is severely impaired, and AIDS patients become more susceptible to what are called "opportunistic" infections such as pneumonia, tuberculosis, and rare forms of cancer.

PHASES OF AIDS

The disease typically goes through three major phases. During the first stage, the individual experiences general flu-like symptoms and remains relatively healthy while the immune system keeps fighting back. However, once the immune system begins to weaken (which may take several years), the next stage includes symptoms like swollen glands, severe fatigue, cough, diarrhea, and night sweats, as well as persistent infections like thrush and herpes. Only in the final stage, which is what is technically called AIDS, does the patient develop the serious infections or tumors that will eventually prove fatal. AIDS is a terribly wasting disease, and its victims are usually reduced to human skeletons as their lungs and brains are also destroyed.

Continued and heavily funded research has not yet found a cure for AIDS, although researchers have discovered drug combinations that effectively prolong the period before full-blown AIDS begins. As with any virus however, HIV has already mutated and developed strains that are resistant to some of these drugs. Research on the origins of AIDS suggests that the virus may be a mutant of a strain that is known to infect the African green monkey, an animal that often comes in contact with humans in West Africa.

To date, information and educational campaigns throughout the world have been effective in making people aware that prevention is, for now, the only sure weapon against AIDS. However, sexual behavior and drug-related activities are not always conducted with common sense in mind. It is especially important to always practice safe sex by using a condom

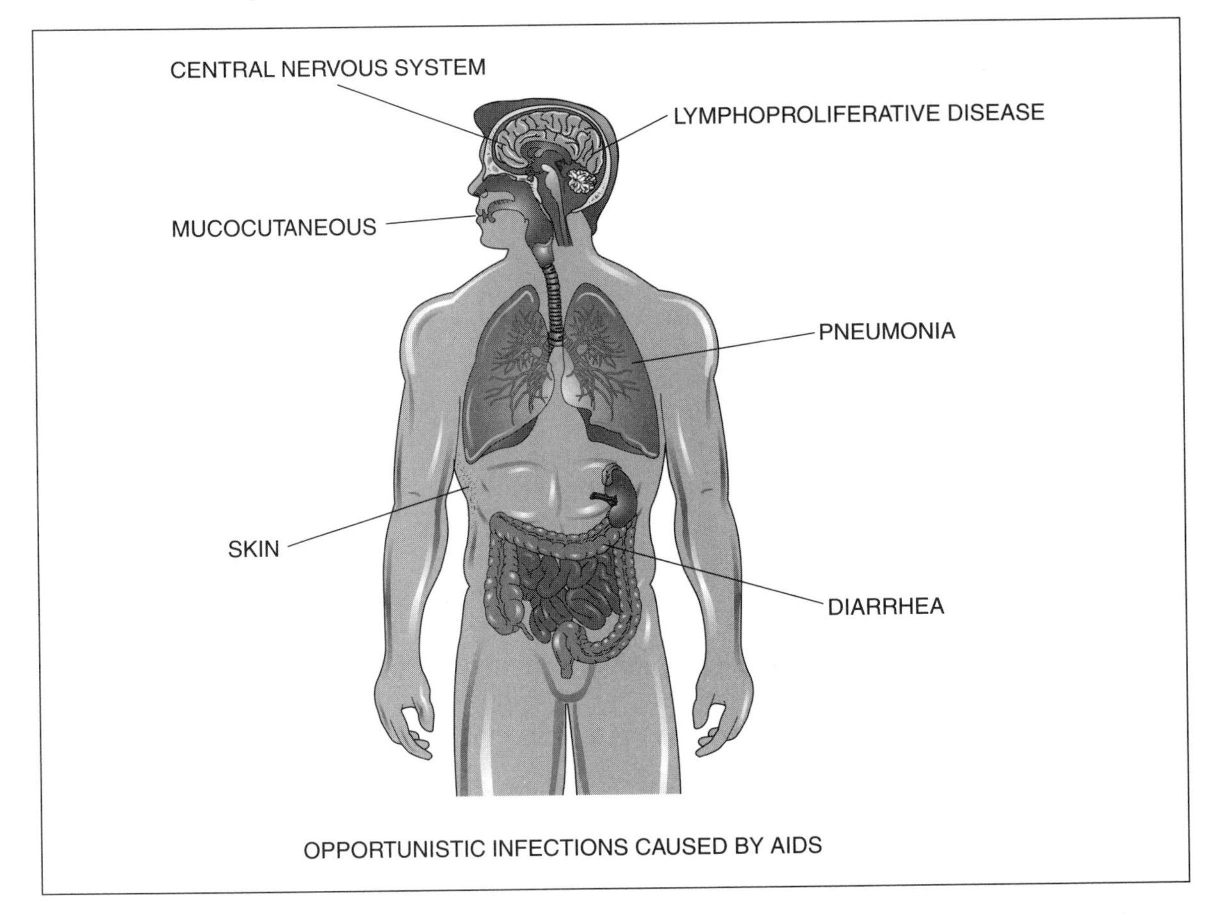

An illustration showing the parts of the body that AIDS attacks and some of the infections that the disease causes. (Illustration by Electronic Illustrators Group.)

(a rubber sheath that is worn over the penis to prevent the exchange of bodily fluids) and not to share intravenous drug needles. Even with widespread awareness, an estimated 6,000,000 new infections occur every year.

[*See also* **Immune System; Virus**]

Algae

Algae are a group of plantlike organisms that make their own food and live wherever there is water, light, and a supply of minerals. Like plants, algae contain chlorophyll, which means that they produce their own food by photosynthesis (using sunlight). There are thousands of different species of algae, ranging from microscopic diatoms to huge ribbons of seaweed. Since algae use the Sun's energy to make their own food, they eventually become food for other kinds of life, just as plants do. Classifying algae is difficult, and experts are constantly revising their ideas. Although algae resemble plants in many ways (they have cell walls and make their own food), some algae can move about and even absorb organic food like animals. However, most algae belong to the kingdom Protista and are considered to be the simplest form of plants (despite the fact that they have no roots, stems, or leaves).

Most algae are aquatic, meaning that they need water to survive, grow, and reproduce. Algae can be found almost anywhere there is water. One of the most common is often seen as the green on the surface of ponds or streams. Some algae merely float near the water's surface while others are capable of moving about on their own. Seaweed is a common form of algae that can be found in the ocean or on the seashore. Even the green, powdery film sometimes seen on trees or on old wooden fences is caused by an algae named Pleurococcus, which is unusual because it is one of the few algae that can survive away from water. Since all algae require light to make food, most are found near the water's surface, although photosynthesis can occur in extremely clear water at a depth of 100 meters (328.1 feet). Marine algae are very important to life on Earth since they produce about 90 percent of the oxygen that is created by the process of photosynthesis. Like green plants, algae take in the carbon dioxide that humans and animals exhale and release the oxygen that humans and animals need to breathe. Single-celled algae make up a large part of the phytoplankton of the oceans. Phytoplankton are found at the beginning of the food chain and form the basis of all nutrition in the sea. Even some whales feed on phytoplankton. Since all marine life is ultimately dependent on this first link in the food chain, there would be no fish to catch and eat without algae.

TYPES OF ALGAE

The different species of algae are grouped into phyla (related groups) according to their pigment (color) and the form in which they store food. Some are better known and easier to identify than others. The six main phyla of algae are diatoms, dinoflagellates, euglenas, green algae, red algae, and brown algae.

Diatoms. Diatoms make up the largest and most important part of phytoplankton and are also the easiest algae to identify (besides seaweed). Each one-celled diatom is protected by a tiny, two-part case that it makes out of silicon dioxide. When a single marine diatom dies, its hard case or shell drifts to the ocean bottom where over time, thick layers of these cases accumulate and are compressed to form a rock called diatomite. This valuable powdery rock is almost pure silica and is used commercially as an abrasive, filtering, or insulating material.

Dinoflagellates. Dinoflagellates almost always live in salt water and form the second most important part of phytoplankton after diatoms. These one-celled algae move about using whiplike tails called "flagella." As algae, dinoflagellates possess chlorophyll but they have red pigment rather than green. When the dinoflagellate population sometimes explodes for unknown reasons, it causes what is known as a "red tide." Red tides sometimes contain a nerve poison that can kill fish and people who eat infected fish.

Euglenas. Euglenas live in fresh water and are able to move with a long, whipping tail. Euglenas combine both plant and animal characteristics. Like plants they are able to produce their own food through photosynthesis. However, like animals, euglenas are also able to capture and eat food. Although they do not have a cell wall, they have a flexible layer inside their membrane as well as an "eyespot" that responds to light.

Green Algae. Green algae make up the phylum Chlorophyta and are distinguished by the presence of chlorophyll. They can be one-celled or many-celled and usually live in water, although they can survive in other environments (like the damp side of a tree trunk). Many species of green algae form colonies (a permanent group of related organisms) or grow in long chains, although some form a ball-shaped colony. Sea lettuce that grows in salt water is a good example of green algae.

Red Algae. Red algae are multicelled and get their name from their distinctive coloring. Since their unique red pigment allows them to absorb even the smallest amount of light, they are able to live far below the ocean surface and still make their own food by photosynthesis. Their food is a type of carbohydrate or starch called carrageenan, and it is used commercially to give toothpaste and even pudding its smooth creaminess.

Brown Algae. Brown algae are multicelled and most species live in salt water. Brown algae, also called kelp, grows into fields that are sometimes 100 yards (91.4 meters) long. Kelp or brown algae play an important role in the ocean as they provide both food for many fish and invertebrates (animals without a backbone) and a place to live and hide for many small fish. People in many parts of the world eat brown algae, and it is used commercially in ice cream, marshmallows, and fertilizer. Despite the fact that some kelp can grow as long as 100 feet (30.48 meters), they lack the complex structure of plants and are still considered algae.

ESSENTIAL TO LIFE ON EARTH

Algae play a key role in sustaining life on Earth since they give off oxygen and absorb much of the carbon dioxide that is produced by not only be humans and animals, but also the burning of fossil fuels. They also form the basis for most food chains in fresh water and ocean habi-

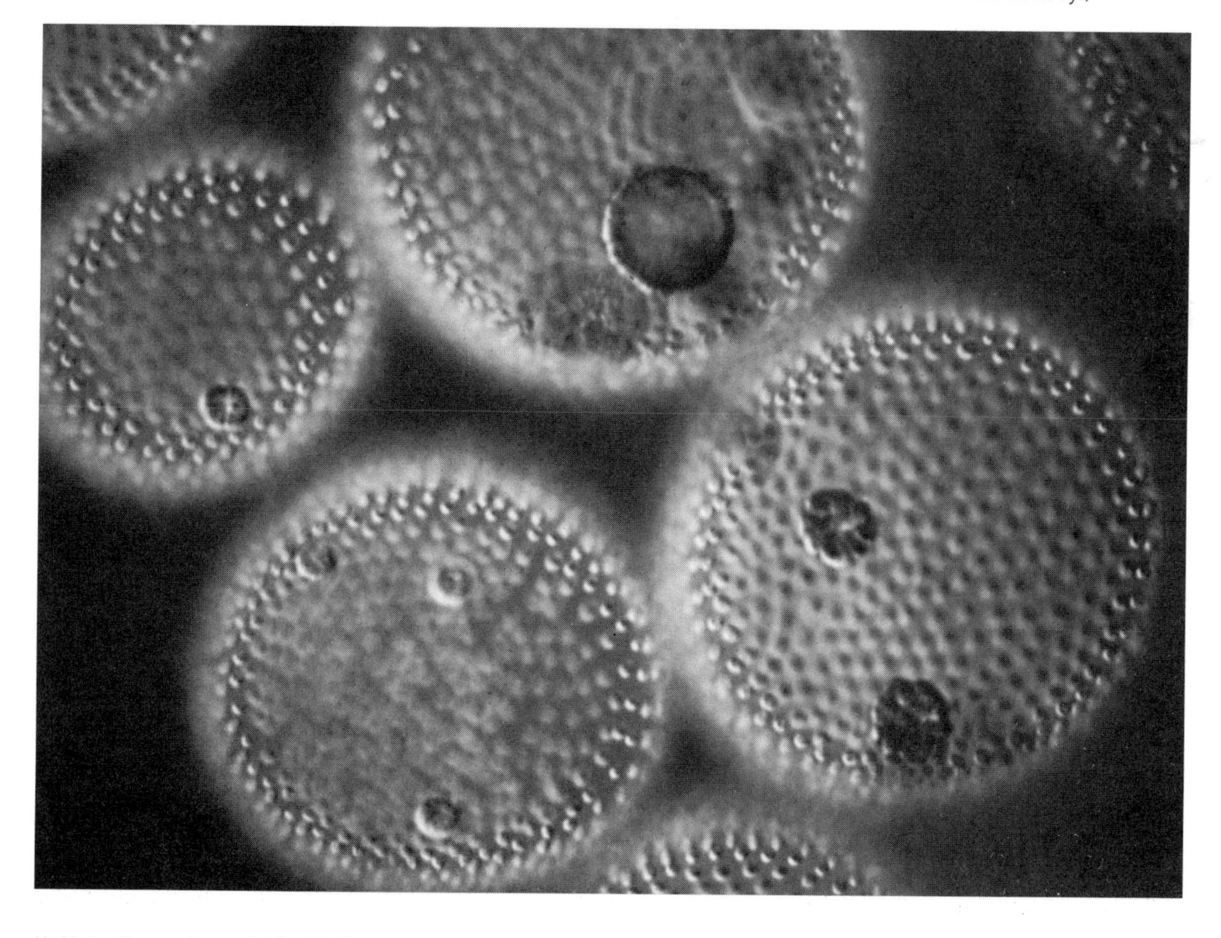

Green algae magnified fifty times its original size. Green algae receives its color from the presence of chlorophyll. (Reproduced by permission of Custom Medical Stock Photo, Inc. Photography by Alex Rakoey.)

tats and have many valuable commercial purposes. For example, brown algae provide a natural source for the manufacture of chemicals called alginates that are used as thickening agents and stabilizers in the industrial preparation of foods and pharmaceutical drugs.

[*See also* **Food Web/Food Chain**]

Amino Acids

Amino acids are the building blocks of proteins. Living things are able to produce vast types of proteins out of only twenty different amino acids. Nearly half the number of amino acids needed to make proteins cannot be made by the body and must be obtained from food.

Proteins are essential to all living things, and for animals especially they perform many critical tasks. Without proteins, there would be no life,

A computer-generated model of glycine, and amino acid. (©Scott Camazine, National Audubon Society Collection/ Photo Researchers, Inc. Reproduced by permission.)

and without amino acids, there would be no proteins. While there are many types of amino acids (some two hundred and fifty have been found in plants and fungi), for humans there are only twenty that are found in proteins. Out of these twenty amino acids, the body is able to manufacture the thousands of different proteins that it needs for growth and repair. The body uses various combinations of the twenty acids to make proteins in much the same way the twenty-five letters of the alphabet are used in various combinations to make different words. Proteins are therefore made of amino acids linked or bonded together in different chainlike combinations. The properties of an individual protein are determined by the sequence or order in which the amino acids are linked together. Therefore, different proteins have different sequences of amino acids.

While most bacteria and plants can make all the amino acids they need, the cells of humans and most other animals are not able to manufacture all of the biologically important amino acids. It is estimated that of the twenty amino acids needed by humans, only ten can be made by the human body. The other ten must be obtained from the food the organism eats. In fact, the human body really gets all of its essential amino acids from the food it consumes since it is only able to make some amino acids by converting the others that it gets from outside its food. Therefore, unless the body obtains certain "essential" amino acids from its protein food source, it does not have the building blocks to make any other amino acids. In such a case, the body's protein-making ability would break down. Nutritionists have determined that all of these essential amino acids can be obtained from meats, eggs, milk, cheese and other foods derived from animals.

[*See also* **Nutrition; Protein**]

Amoeba

An amoeba is a single-celled organism that has no fixed shape. As a protozoan and a member of the kingdom Protista that has animal-like qualities, an amoeba has to find and eat its food (since it is unable to make its own food as plantlike Protists do). Of the many species of amoeba, some live in water and soil, while others are parasites and live inside plants or animals.

Many biologists believe that the first protozoans were similar to today's amoeba. As members of the phylum (or primary division of the animal kingdom) Sarcodina, amoebas are distinct among protozoa in that they have no definite shape. Although they consist of a single cell surrounded by a membrane, they move by changing the shape of that membrane and therefore do not have one particular shape that makes them immediately recognizable. Some would say that they are recognizable because of this formlessness, while others say they look like a tiny bag of jelly. An amoeba actually moves in a very strange manner using pseudopods or "false feet."

When it wants to move, an amoeba turns its jelly-like body solid in a certain spot to form a temporary "foot" which it stretches in the direction it wants to move. The rest of the amoeba then flows into the pseudopod and basically changes its position to where its pseudopod had reached. This type of movement is called "streaming." Every time an amoeba does this it changes its shape. Since this is not the fastest way to move about, an amoeba only can move at a top speed of about 1 inch (2.54 centimeters) an hour.

An amoeba eats its food in much the same manner as it moves. It flows around and surrounds the food and then the amoeba totally engulfs it. The food is then held in a food vacuole, which is a specialized structure that digests the food. The amoeba mainly eats bacteria, algae, and other protozoans. Any waste that remains after the food is digested is released from a contractile vacuole (one that can open and close). Water

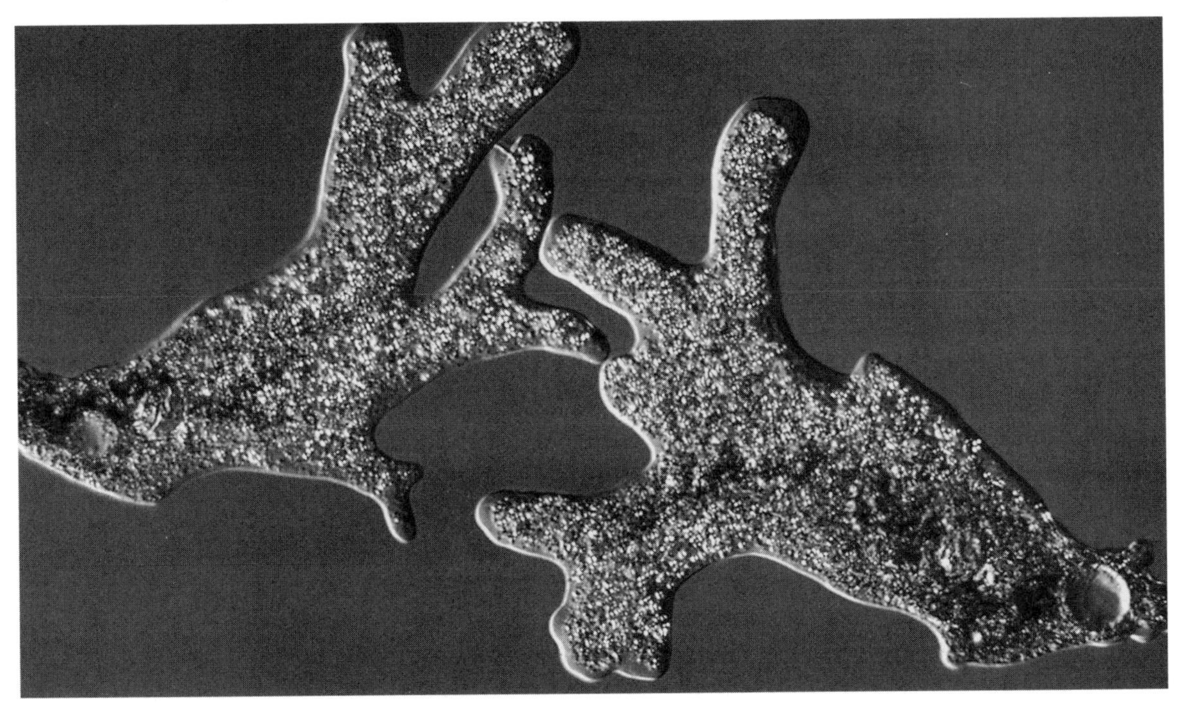

Amoebas are formless organisms that use pseudopods, or "false feet" to move about. The false feet are visible on the amoeba in this photo. (Reproduced by permission of Photo Researchers, Inc. Photograph by M.I. Walker.)

also flows into the amoeba by the process of osmosis (in which water flows through a membrane until the solutions on either side of it are at equal strength). When the amoeba needs to expel some water, it squeezes its contractile vacuole and squirts some out.

There are four types of protozoans that are called amoeboid protozoans. All use pseudopods to move about and to capture their food, and three of them are considered shelled amoeba since they are at least partially covered by a shell made up of the minerals they secrete. The first of the amoeba is the familiar "naked" amoeba called *Amoeba proteus.* While most of this type of amoeba live in water, some are parasitic and live in the human gut. The infected human then contracts the disease called amoebic dysentery. One of the shelled amoeba that lives in a tiny chambered sea shell and uses threadlike pseudopods to move about is called a foraminiferan. This amoeba can barely be seen with the naked eye. A second shelled amoeba is the freshwater heliozoan or "sun animal." It has thin pseudopods that look like needles that radiate from its body like rays from the Sun. The third and most intricate of the shelled amoebas is the radiolarian. These sea creatures have skeletons of silica, a mineral that does not dissolve in deep water, so that the deepest seabeds are covered with a thick layer of what is called radiolarian ooze.

Amoebas usually reproduce asexually by a process known as binary fission. In this form of reproduction, an amoeba splits in two after pinching in half and forms two smaller but identical cells. This occurs after the cell's nucleus duplicates its hereditary material and divides in two.

[*See also* **Cell; Protozoa; Reproduction, Asexual**]

Amphibian

An amphibian is a cold-blooded vertebrate (an animal with a backbone) animal that spends part of its life in the water and part on land. After hatching from an egg, an amphibian usually lives in water and breathes through gills. As it grows, it undergoes a metamorphosis, growing legs and developing air-breathing lungs. At home in both water and land, the amphibian lives in damp places where its thin skin will not dry out.

THE LIFE CYCLE OF AN AMPHIBIAN

The name amphibian comes from Greek words meaning "having two lives," and it is this unique life cycle that most characterizes an amphibian. Frogs, toads, newts, and salamanders are the best-known members of this smallest class of vertebrates. They all share several other characteristics. Most have thin, moist skin that is smooth and soft. They do not have scales or claws, although the skin of a toad is dry and covered with bumps. They have a three-chambered heart (a fish has a two-chambered heart and birds and mammals have four-chambers) and are ectothermic or cold-blooded. This does not mean that they are always cold, but rather that their body temperature matches that of their surroundings. Because of this, they become sluggish or inactive in the cold and usually hibernate (an inactive state resembling deep sleep) during extreme cold. Most adult amphibians have four limbs and are carnivores (meat-eaters) who will eat almost anything they can catch. Although adult amphibians can live on land, they must return to water to reproduce. There they lay their eggs, which are fertilized by the male outside of the female's body. The jelly-coated eggs remain in water until they hatch into larvae (the early stage of an organism's development, which changes structurally as it becomes an adult).

Metamorphosis. Upon hatching, a process called metamorphosis (a series of distinct changes in form through which an organism passes as it develops from and egg to an adult) begins that is unique among vertebrates. After the eggs hatch in the spring, legless tadpoles emerge. These animals look like tiny fish with long tails, and they breathe as fish do, using their external gills. Soon, however, the tadpoles begin to change gradually as their tails shrink and the beginnings of legs start to form. Covers over their gills start to grow as their lungs begin to take shape. It may take as long as two years for a frog larva or tadpole to change completely into an adult, but when it does, it looks entirely different from when it was hatched. As an adult, the frog has a tail-less, squat, compact body with four legs, the back two of which have powerful muscles for jumping. Upon maturity, the frog will mate and the cycle will begin all over again with its offspring.

HIBERNATION

During the winter months, frogs and toads become inactive since their body temperature decreases. It is at this time that frogs bury themselves in the mud at the bottom of lakes or ponds and hibernate. Toads do the same only in soft, moist soil. During hibernation, their body processes slow down considerably, and they are able to exchange gases (breathe) through their damp skin. Besides frogs and toads, the other major group of amphibians is made up of the tailed amphibians like the salamander and the newt. Both have long bodies, four short, thin legs, and a tail. Salamanders spend most of their adult lives on land, while newts live in the water. As amphibians, they both begin life with gills living underwater, and soon develop lungs as they mature and change into adults.

AMPHIBIANS OFFER GREAT VARIATION

Even in a small class like the amphibians, there is amazing variation. For example, some frogs and toads are poisonous. Tree frogs have toes with sticky pads for gripping, while other frogs are able to jump and then glide using their webbed feet as a parachute. The pink salamander is able to reproduce before it even matures, and the Caecilian is an amphibian without any legs at all, having a wormlike body and scales. All, however, have the double life of an amphibian. Recently, biologists have noticed the disappearance of some species and an overall decrease in the amphibian population. Some believe that the health of amphibians may serve as an early warning system for the overall health of the environment. Since amphibians metabolize toxic substances in much the same way that hu-

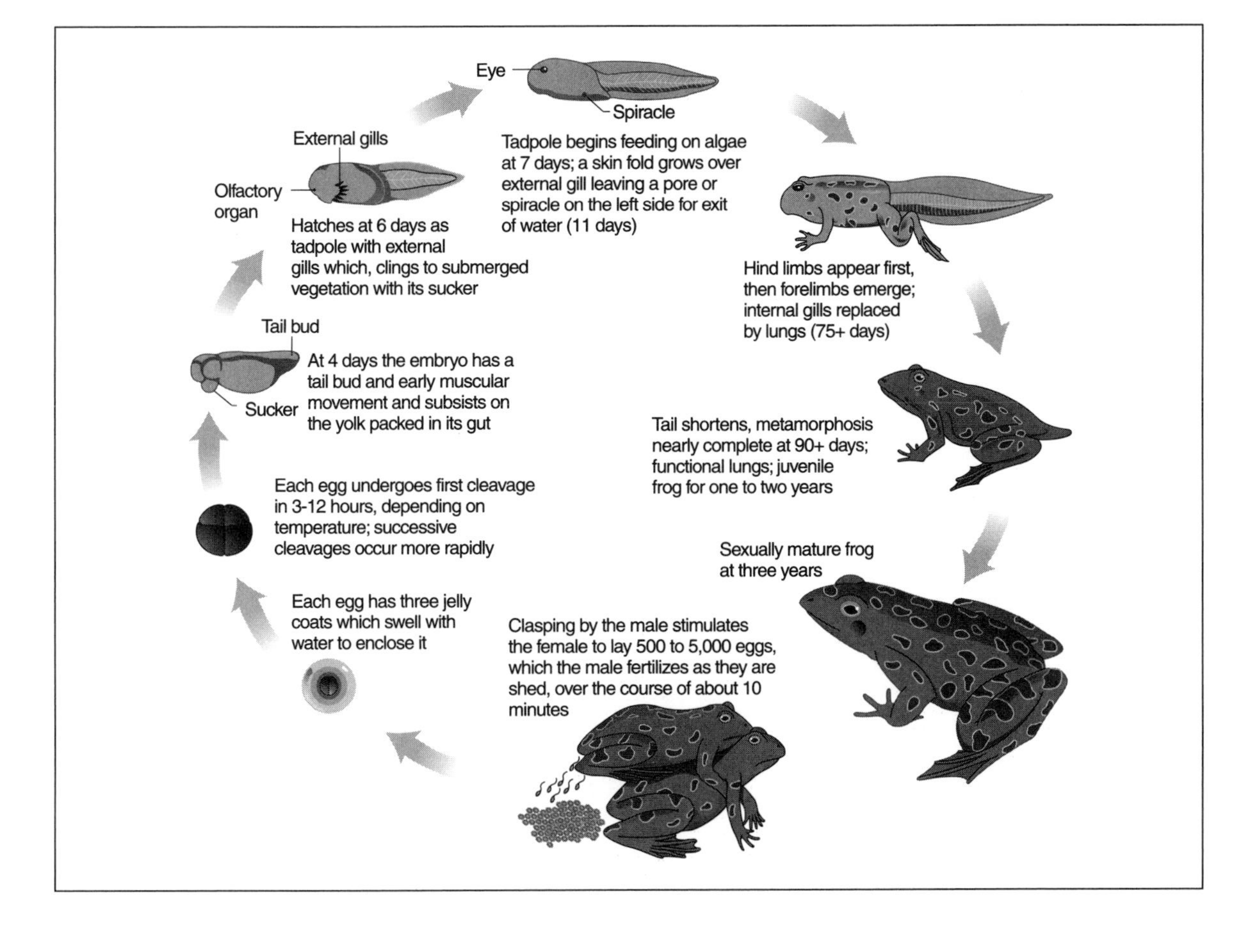

A flow chart of the life cycle of a frog beginning with the fertilization of the eggs and ending with a sexually mature frog. (Illustration by Hans & Cassidy. Courtesy of Gale Research.)

mans do, whenever any serious changes are noticed, such as a sharp drop in population or mutant frogs being born, it suggests that their environment might be equally hazardous to humans.

[*See also* **Animals; Metamorphosis; Vertebrates**]

Anatomy

Anatomy is the study of the biological structure of living things. Although many think of anatomy as being concerned only with the human body, the word actually applies to the structure of plants, animals, and other organisms.

ARTISTOTLE BECOMES THE FIRST TO STUDY ANATOMY

As one of the oldest branches of biology (the science of life processes and living organisms), anatomy comes from the Greek word *anatome,* which means "cutting up." Until modern times, dissecting or "cutting up" was in fact the only way to learn how living things were actually put together. A knowledge of anatomy was important even in ancient times, since it was recognized early on that it was impossible to understand how the parts of a living thing worked until one knew how they were shaped and how they all fit together. The Greek philosopher and scientist, Aristotle (384–322 B.C.), is considered the first to study anatomy, and he is credited with the idea that each organ has its own function that could be discovered by observing its structure. Because of Aristotle's work, the structure and function of a body's organs and parts have been linked from very early times. The Greek scholar, Herophilus of Chalcedon (335–280 B.C.) is credited with founding the first school of anatomy and is believed to have conducted some six hundred dissections. By the first century A.D., dissection of human corpses was becoming discouraged, so the prominent Greek physician, Galen (129–200) dissected apes, dogs, and pigs in order to study anatomy. Galen is considered to be the founder of experimental medicine. Many of his anatomical teachings about the human body were incorrect (as they were based on the anatomy of other animals), but his work was considered to be the final authority for centuries.

VESALIUS BEGINS THE MODERN ERA OF ANATOMY

The modern era of anatomy began with the Flemish physician, Andreas Vesalius (1514–1564), who published his classic *On the Structure of the Human Body* in 1543. His highly accurate drawings were based on his extensive dissection of the bodies of executed criminals. Despite being condemned in his lifetime because his writings contradicted the old teachings, Vesalius is now considered to be the founder of modern anatomy. By the time the great English physician William Harvey (1578–1657) correctly described the circulation of the blood in 1628, the study of anatomy had been reestablished in schools and human dissection was once again permissible.

Anatomical study was broadened with the seventeenth-century invention of the microscope, since it opened up an entirely new world that was too small for the naked eye to see. Regular microscopic discoveries eventually led to the nineteenth-century finding that all living matter is made up of cells. In that century, the second great revolution in anatomy took place when the English naturalist, Charles Darwin (1809–1882), introduced his theory of evolution in 1859. His argument that all living species are descended from other species led to the branch of anatomy called comparative anatomy. Comparative anatomy is used to study the anatomical differences and similarities between animals, and eventually provided evidence for Darwin's theory. Today, the study of anatomy has shifted from the invasive techniques, which usually opened the body in some manner in order to observe, to noninvasive techniques, such as x rays, computerized tomography (CT scan), ultrasound, and magnetic resonance imaging (MRI).

The modern age of anatomy began with Andreas Vesalius and the publication of his highly accurate drawings based on the human body. (Courtesy of The Library of Congress.)

ANATOMY OF ANIMALS AND PLANTS

The anatomy of higher animals is made up of eleven body systems: the integumentary system (external features that are related to its skin); the skeletal system (external or internal); the muscular system (including three different types); the nervous system (including the brain, sense organs, and nerves); the digestive, circulatory, and respiratory systems (that work together to nourish the body); the excretory system (that rids the body of waste); the reproductive system (that allows new life to be created); the endocrine system (that produces hormones that regulate bodily functions); and the immune system (that protects the body from infection).

The anatomy of plants is much simpler than that of animals, having only two main systems—the root system and the shoot system. The root system anchors the plant in the ground and al-

ANDREAS VESALIUS

Flemish anatomist (a person who studies the structure of human and animal bodies) Andreas Vesalius (1514–1564) began the modern era of anatomy with the publication of the first accurate book on the human body. Often called the founder of modern anatomy, Vesalius performed many autopsies (examinations of dead bodies to determine the cause of death) and discovered that much of what was taught about the human body was wrong. The illustrations in his book, *On the Structure of the Human Body*, are at the highest level of both art and science, and it is considered one of the greatest biology books ever written.

Andreas Vesalius was born in Brussels, Belgium, and came from a long line of physicians. Although his mother was English, his father was court pharmacist for Emperor Charles V, and Andreas studied medicine in Belgium and France. At that time, medical schools were very conservative in that their teachings were based on very old texts written around A.D. 175 by the Greek physician, Galen (A.D. c.130–c.200). Galen wrote about anatomy, which is the study of the physical structure of living things, and specialized in human anatomy. However, when Galen did his work, it was unlawful to dissect (to cut open and examine) the bodies of dead people, so Galen did most of his anatomical research on dead animals like monkeys, pigs, dogs, and goats. While he did advance the study of anatomy with this work, not all of it was directly applicable to the human body. Nonetheless, some 1500 years later, Galen's teachings were still being used in many of the more conservative medical schools.

Vesalius had an inquiring mind, and when he thought Galen was wrong he told his teachers. This only served to get him in trouble, so he moved to Padua, Italy, and earned his medical degree from that city in 1537. Although dissecting humans was still discouraged, things were much freer in Italy,

lows it to get water and nutrients from the soil. The shoot system is made up of all aboveground stems, branches, leaves, and flowers.

In the life sciences, the study of anatomy is essential to our understanding of the overall structure of living things and of how those individual parts relate to one another, influence one another, and work together. Understanding the anatomical similarities of different organisms provides important evidence of how all living things are linked together through the process of evolution.

[*See also* **Circulatory System; Digestive System; Endocrine System; Muscular System; Nervous System; Reproductive System; Respiratory System; Skeletal System**]

and when Vesalius began to teach anatomy himself, he did something that was truly revolutionary. It was the practice for teachers to lecture during dissections and only supervise the actual cutting, which was done by assistants who usually knew little about the human body. Vesalius, however, decided to perform these dissections himself as he taught, and his lectures became popular with the best students. Finally, he realized just how wrong much of Galen's teachings were, and he decided to produce an anatomical textbook that, once and for all, would actually show the way the body was really constructed. He commissioned talented artists to draw the anatomical features of the human body the way he actually saw them when he dissected. One story tells how, until Vesalius, it was taught that men had one fewer rib than women because of the creation story in the Bible. Vesalius put nothing in his book that he had not observed himself, and after three years of hard work, he published his *De humani corporis fabrica* (*On the Structure of the Human Body*), whose highly accurate and artistically beautiful woodcuts raised anatomy to a new level. The publication of this great work instantly marked the beginning of modern anatomy and introduced a new standard for anatomical textbooks. Today it is considered to be one of the greatest medical works ever produced. Yet in its time, it was actually ridiculed by the medical establishment.

Although this work would eventually revolutionize biology, it would bring Vesalius as much trouble as it did fame. His enemies accused him of snatching bodies to dissect, and he was even accused of religious heresy (having an opinion in opposition to religious beliefs). At one point, Vesalius became so disgusted that he gave up his work altogether. Eventually, he was given a good position at the royal court, but was ordered to make a pilgrimage, or journey, to the Holy Land (the Middle East) to make up for his heresies. It was during his return trip that he died off the coast of Greece when his ship was wrecked in a storm.

Animals

Animals are a group of multicelled, living organisms that take in food. Most animals reproduce sexually (with sperm fertilizing an egg), can move about, and are able to respond to their surroundings. Of the separate divisions of living organisms, the animal kingdom forms the largest in terms of the number of species.

Animals range in size from barely visible one-celled animals to the 100-foot (3.48 meter) blue whale. Different creatures may slither, burrow, climb, run, swim, and fly—yet they are all considered part of the Animalia kingdom. The Animalia kingdom is one of the five major divisions of all living organisms. The other kingdoms are Monera (bacteria with no nucleus); Protista (one-celled organisms with a nucleus); Fungi (multicelled organisms that take in food); and Plantae (multicelled plants that make their own food).

CHARACTERISTICS OF ANIMALS

There are six major characteristics of all animals, whether they are worms or whales. First, animals cannot produce their own food (as plants do) and must therefore rely on eating other living things. Second, animals cannot use protein, fats, and carbohydrates directly and so must first digest or break them down into smaller molecules. Third, because of their need to find food and a mate, as well as escape from an enemy, animals have developed a way to move from place to place. Fourth, animals are multicellular, having many cells that are highly specialized. Fifth, animal cells are eukaryotic, meaning that each has a nucleus surrounded by a membrane. Sixth, animals are able to respond quickly and in the correct manner to changes in their environment.

VERTEBRATES AND INVERTEBRATES

To classify the different types of animals or to group them by their similarities, biologists have divided animals into vertebrates (those with a backbone) and invertebrates (those without backbones). Although vertebrates, such human beings, whales, elephants, and dolphins, are the biggest and brainiest of the animals, about 97 percent of the entire animal kingdom is made up of invertebrates like worms, sponges, clams, and insects. The next thing a person who classifies animals considers is the physical arrangement of an animal's body parts. Some animals (like humans) have what is called "bilateral symmetry." This means that if an imaginary line were drawn from top to bottom of an animal, each half of its body would be a mirror image of the other. Those with "radial symmetry," like sea anemones, have their body parts arranged in a circle around a central point. Others like a sponge, have no definite shape at all and are called "asymmetrical."

IDENTIFYING BY FEATURES

Among the major divisions of better-known animals, it is possible to group several according to some of their easily identifiable features. One of these groups includes animals with exoskeletons or strong and light skeletons on the outside of their bodies. Examples of animals with exoskeletons include horseshoe crabs, oysters, and snails as well as insects like spiders and ticks. Animals with a backbone include fishes with gills, amphibians who live on land but need to breed in water, and cold-blooded (animals whose body temperature changes with the environment) reptiles like lizards, snakes, and crocodiles. Birds evolved from reptiles as their scales changed into feathers. They lay eggs and can fly. Mammals are warm-blooded and give birth to live young. The strangest group of animals may be the echinoderms or "spiky skin" animals like starfish. Their structure and shape are completely different from other animals, including five identical parts and a skeleton of plates inside their bodies. Although echinoderms have no head, brain, or blood, they are still members of the animal kingdom. The animal kingdom, of which human beings are one part, includes a wide variety of different but related life forms.

[*See also* **Arthropods; Echinoderms; Mollusks;**]

Antibiotic

An antibiotic is a naturally occurring chemical that kills or inhibits the growth of bacteria. Today, antibiotics are used to treat infections and to fight a wide range of bacteria. However, the overuse of antibiotics has caused bacteria that are resistant to antibiotics to become more widespread, and in many cases, the antibiotics have become ineffective.

When a person gets an infection, microscopic bacteria have entered the body through an opening or a wound. After quickly finding an abundant supply of food inside, these bacteria reproduce in great numbers and release toxins or poisons as they grow. These toxins can interfere with cell functions or even destroy human cells.

THE HISTORY OF ANTIBIOTICS

Antibiotic drugs have been developed to fight and kill bacteria. They are derived from other organisms, like molds, that are naturally harmful to bacteria. Certain molds produce their own toxins that destroy bacterial cells. This may be the means by which a mold would defend itself against bacterial invasion. As early as 1871, the English surgeon Joseph Lister (1827–1912) noted that certain organic compounds seemed to act against bacteria. However, it was not until 1928 that the Scottish doctor Alexander Fleming (1881–1955) made the important discovery that would eventually lead to the development of penicillin (synthetically produced antibiotics derived from molds and used to treat a wide variety of diseases). While he was growing cultures of bacteria in petri dishes for experiments, Fleming accidentally left several dishes uncovered for a few days. He then noticed that a green mold had gotten into one dish (having traveled through the air as a mold spore) and had destroyed or dissolved the bacteria. Examining the situation with his trained eye, Fleming realized that he had come upon a natural substance that could kill bacteria. His later experiments with mice showed that his new "penicillin" killed only the bacteria and did not harm the animals' cells. Since he was unable to purify and concentrate more penicillin, he published a paper that received little attention. It was not until 1940 that penicillin was taken up experimentally by others who, by 1942, were beginning to make it in large amounts. Fleming's discovery would eventually lead to the steady production of several different lifesaving antibiotics.

HOW ANTIBIOTICS WORK

The key to why an antibiotic works is that it is selectively toxic or poisonous. That is, it works against certain life forms and not others. It does this by interfering with the cell wall of each new bacterial cell, and this eventually kills the cell. Since animal and human cells do not have cell walls, it is not harmful to these types of cells. However, when an antibiotic encounters a bacterial cell, it joins with its cell wall, leaving a gap in the cell wall so that it no longer can protect its contents, which then spill out. Other antibiotics bind to the ribosomes (particles that act in protein synthesis) in a bacterial cell and stop them from making proteins (which a cell needs to stay alive).

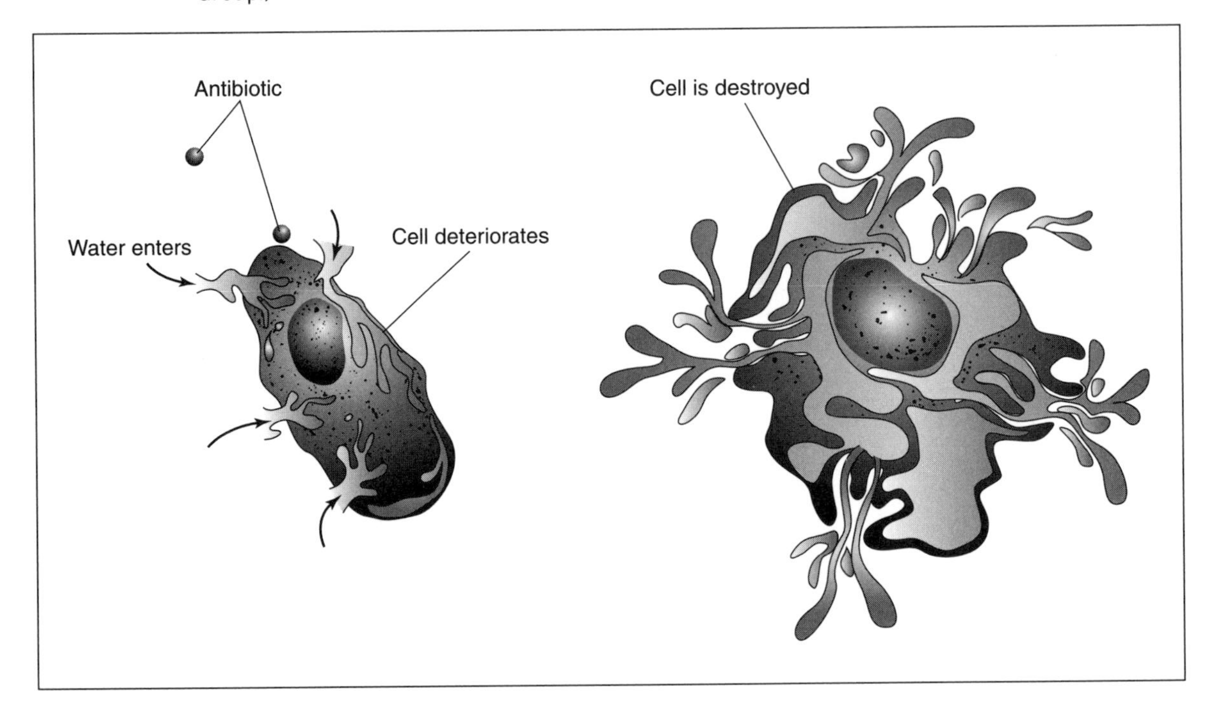

A labeled illustration showing how antibiotics attack and destroy a bacterial cell. (Illustration by Electronic Illustrators Group.)

BACTERIOCIDAL AND BACTERIOSTATIC ANTIBIOTICS

There are two types of antibiotics: some antibiotics are bacteriocidal while others are bacteriostatic. Bacteriocidal agents kill bacteria, while bacteriostatic agents slow them down so that the host's immune system has a better chance to defeat them. Today's many antibiotics can be broad-spectrum agents or narrow-spectrum agents. As it sounds, one is effective against a broad range of bacteria while the other works against only a few. There also are many different "families" of antibiotics, some of which are synthetic or man-made. However, no matter how different they are from one another, when a patient is given an antibiotic injection or takes an antibiotic pill, the antibiotic prevents bacterial cells from growing and dividing normally.

ANTIBIOTIC RESISTANT BACTERIA

Bacteria may seem easy to kill with modern medicine. Yet these invisible agents of disease can reproduce every twenty minutes and have proven capable of becoming resistant to antibiotics. They do this by mutations, or changes. that occur in a cell's genetic material. More and more, antibiotic resistant bacteria are becoming increasingly common due largely, it has been shown, to their routine use by farmers who give antibiotics to their livestock and chickens to prevent them from getting sick. Unfortunately, because people then consume the meat of these animals, humans are ingesting some of the antibiotics given to the animals. It has been shown that overuse of antibiotics in both humans and in animals speeds up the development of antibiotic resistant bacteria.

Without antibiotics, humans still would be subject to the terrible diseases that killed millions of people in the past. Before 1950, bacterial diseases like diphtheria, tuberculosis, pneumonia, blood poisoning, food poisoning, bacterial meningitis, and scarlet fever were sure killers. This victory over bacteria may be coming to an end since humans are now faced with the real possibility that entire populations of bacteria are mutating to the point where they will be resistant to any antibiotic available.

Antibody and Antigen

An antigen is any foreign substance in the body that stimulates the immune system to action. An antibody is a protein made by the body that locks on, or marks, a particular type of antigen so that it can be destroyed by other cells. Antibodies are an essential part of the immune system of vertebrates (animals with a backbone) and enable the body to resist disease-causing organisms.

All vertebrates have an immune system that produces antibodies. The immune system is able to distinguish "self" from "nonself" and recognizes when an antigen (foreign cell) has invaded the body. The immune system then produces special chemicals called antibodies to fight and help kill the invader. The immune system is also able to "remember" these specific invaders, and if they ever return, it is able to respond even faster using specific antibodies whose job is to lock on to that particular type of antigen. After locking on, or binding with it, the antibodies get help from other cells and proteins that destroy the antigen or at least neutralize it.

Antibodies really work after the fact. When an antigen, such as a virus, invades the body, two things can happen. However, if this invader is new to the body and has never entered it before, the body has no antibodies to combat it. In such a case, it is the body's large white blood cells known as macrophages that will attack and try to destroy the antigen. If this virus has entered the host before, then specific antibodies already exist. These antibodies will immediately recognize the virus as "nonself" and bind to it like a key in a lock. Once they lock on to the antigen, they have marked the invader as a target for the body's killer cells. An antibody will only recognize and help destroy one kind of organism or antigen. If a new and different organism enters the body, a new type of antibody must be produced.

Although scientists were aware of antibodies in the 1890s, it was not until the late 1930s that scientists came to discover what they really were. In 1938 antibodies were identified as proteins of the gamma globulin portion of the plasma (the liquid portion of the blood). Later it was found that antibodies are produced by special white blood cells called B-lymphocytes.

Immunization, sometimes called vaccination, uses the ability of the immune system to remember a previous invader. For example, a child is immunized against certain diseases, like measles, mumps, rubella, diphtheria, whooping cough, tetanus, polio, and chicken pox, through a vaccine. Vaccines contain dead or weakened disease-causing organisms that stimulate the body's immune system without actually causing the disease. Before vaccination, these diseases were common among children and responsible for many deaths. Now, routine vaccination of children has virtually eliminated these diseases. Vaccination works because once a certain antibody is produced in the body, it usually remains for many years.

The case of immunization is an excellent example of how an understanding of the body's systems and operations allows scientists to better use the body's own natural defense mechanisms to the advantage of the individual.

[*See also* **Blood; Blood Types; Immune System; Immunization; Rh Factor**]

Arachnid

An arachnid is an invertebrate (an animal without a backbone) that has four pairs of jointed walking legs. Most arachnids, like spiders, ticks, scorpions, and mites, live on land and have two main body parts. Many arachnids prey on other invertebrates, while some are parasites. Unlike insects, however, arachnids have no jaws, antennae, wings, or compound eyes

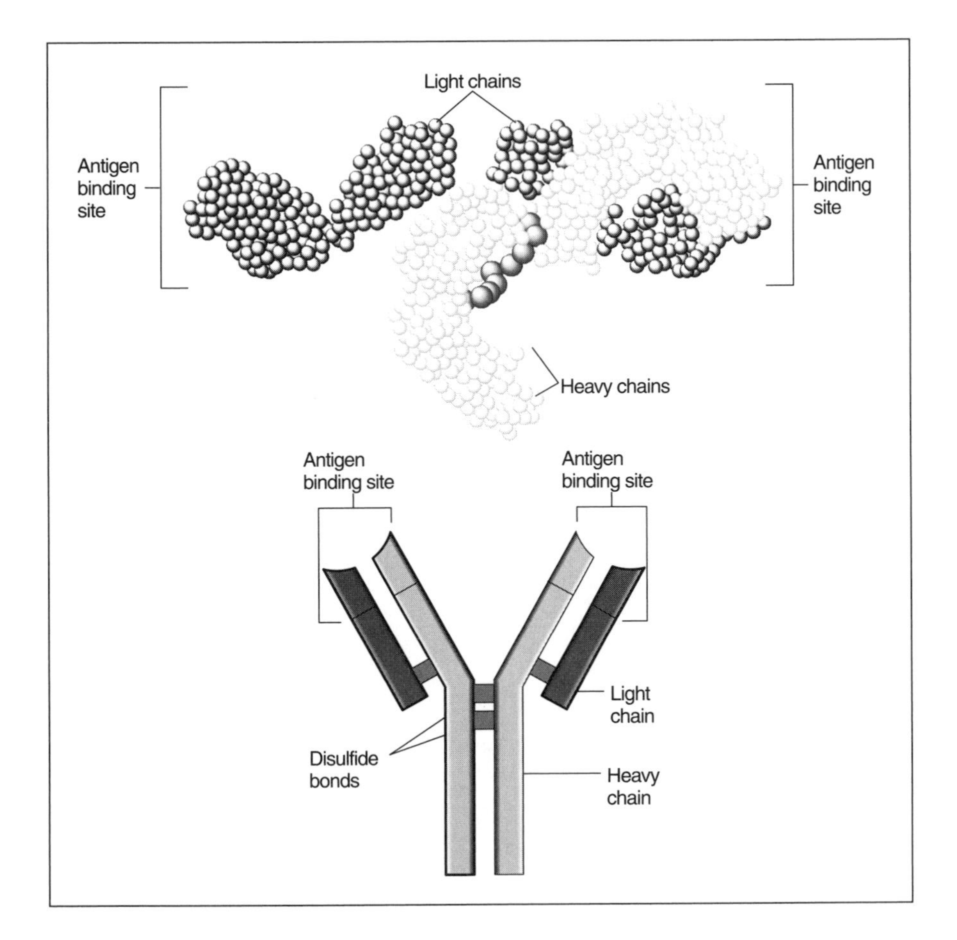

Two illustrations of the molecular structure of antigens. (Illustration by Hans & Cassidy. Courtesy of Gale Research.)

(many tiny, individual eyes that altogether make a single image that resembles a mosaic). Arachnids are probably the most unpopular type of arthropod (an invertebrate with jointed appendages and a segmented body).

An arachnid is a member of the phylum (or primary division of the animal kingdom) Arthropoda since it has a jointed body case, an exoskeleton (a hard, outer support structure), and no backbone. The class name Arachnida comes from a story in Greek mythology in which a young girl named Arachne spun a silken cloth more perfectly than did the goddess Athena, who became so angry that she turned the girl into a spider. Arachnids are sometimes thought to be insects, but what defines an arachnid and prevents it from being considered an insect is the number of legs it has. An arachnid has eight legs. An insect has six legs. A crustacean has ten legs. Thus a spider and a tick are arachnids, while a centipede and an ant are not. Among other characteristics of arachnids is their highly developed sense of sight. They also have a rigid yet versatile exoskeleton that, combined with muscles and jointed limbs, allows them to have great flexibility and mobility. Its skeleton armor also keeps them waterproof and prevents them from losing body fluids by evaporation.

The spider is a good representative of the arachnid class. Besides their eight walking legs, they have a modified pair they use for handling food and another modified pair of appendages they use as poison fangs or claws.

A spider spinning a web. Spiders are probably the best example of the arachnid class. (Reproduced by permission of JLM Visuals.)

Most spiders breathe by means of "book lungs." These are special organs located on the abdomen; they include chambers filled with hollow plates connected to tubes that extend to the outside of the skeleton. Air enters via these tubes and passes over plates that are richly supplied with blood. The oxygen diffuses into the blood and carbon dioxide is released. Arachnids, like spiders, are usually hunters, so they have sense detectors in their front end as well as six or eight pairs of eyes. Web-weaving in order to trap prey is probably the most distinctive characteristic of a spider, although not all spiders spin sticky webs. Some literally jump on top of their prey. Weaving a web is instinctive behavior, and webs are made of liquid silk (really a protein) produced from glands in the abdomen and shot out through "spinnerets" at the rear of spiders' bodies. Sticky threads are added later to catch and hold any prey that gets tangled. Spiders wait in hiding and stay connected to their web by a single thread that acts like a fishing bobber and signals that something is caught. The spider's feet are coated with an oily film so that it will not be caught in its own web. The spider bites its prey, immobilizing or killing it, and wraps it in silk for a later meal. Other arachnids, like scorpions, simply paralyze their prey with a sting. Similar to a spider, scorpions consume only the prey's bodily fluids.

Unlike spiders and scorpions, ticks are parasites. They burrow their heads into the skin of a mammal and feed on its blood. A tick is able to take in 200 times its own weight in blood at a single feeding. Although they do not always kill their victims, ticks may sometimes carry infectious diseases, such as lyme disease, which may be fatal if left untreated.

Arachnids reproduce sexually (through the union of male sperm and female eggs) and therefore must mate to produce offspring. For spiders however, mating can be a dangerous act. Since spiders not only eat other spiders but also those of their own species, it is not unusual for the female to kill and eat the smaller male immediately after mating with him. To avoid being eaten, some male spiders lock fangs with the female while they mate. Others tie the female in silk before mating, and still others present her with a meal already wrapped and ready to be eaten. Females lay their eggs in silken cocoons and sometimes keep watch over them until they hatch.

[*See also* **Arthropods**]

Arthropod

An arthropod is an invertebrate (an animal without a backbone) that has jointed legs and a segmented body. Arthropods are the world's oldest creatures and the most successful invertebrate group. The phylum Arthropoda contains nearly 1,000,000 species and is therefore the largest in the kingdom Animalia. It is estimated that arthropods account for 75 percent of all the animals on Earth. They range from dust mites to huge crabs.

EXTERNAL STRUCTURE OF ARTHROPODS

The name arthropod is translated as "jointed foot" but it is really a jointed or segmented body that is most characteristic of an arthropod. Since all arthropods have an exoskeleton, which is a hard covering that surrounds the outside of an animal's body, their outer skeleton cannot consist of a single solid piece or they would never be able to move about with any flexibility or speed. A tough covering that is composed of overlapping plates (linked together by tough but flexible hinge joints) allows the arthropod to bend, twist, and move about with great freedom. The same applies to any appendages (legs, arms, tails, pincers) an arthropod may have; these too are made up of tough but flexible joints that allow them to maneuver easily over most surfaces. Whether a lobster or a beetle, most of the arthropod body is covered by an exoskeleton called a cuticle. This protects its soft tissues from predators and disease and supports its entire body. The cuticle is made of a protein called anthropodin, and a carbohydrate called chitin that together produce a tough and flexible covering. This exoskeleton can vary immensely. It may consist of the delicate and flexible wing of an insect or to the heavy and thick shell of a lobster. The major drawback of an exoskeleton is that it makes growth or physical expansion difficult. Since chitin is not living tissue, it cannot expand, and must instead be shed and regrown when an animal's body gets larger. This periodic shedding is called molting and occurs when an arthropod splits its exoskeleton and walks out of it to later form another. During this stage an arthropod is especially vulnerable to attack. The soft-shelled crabs that people eat and enjoy are actually hard-shelled crabs that are caught while molting.

INTERNAL STRUCTURE OF ARTHROPODS

Inside their suit of armor, all arthropods are basically the same. All have a nervous system made up of a brain, simple eyes, and nerves that connect to a long nerve cord running the length of their body. This allows them to perceive and react to their environment. Arthropods that live on land breathe through a tracheal system rather than through lungs. This system consists of narrow, air-filled tubes in their outer skeletons called tracheae that branch into smaller tubes inside the body and directly supply each cell with the oxygen it needs. Water-living arthropods breath with gills, located sometimes on their bodies and other times on their legs. Their circulatory system is an open one, meaning that blood flows freely throughout their body, pumped by a simple heart. Digestion always begins with a mouth and continues into a single long gut that runs the length of their body. Excretion of waste takes place from a separate opening. Most arthropods reproduce sexually (through the union of male sperm and female eggs), although a few species have both male and female organs. Most females lay their eggs in a protected place where they eventually develop into larvae.

TYPES OF ARTHROPODS

Crustaceans. Arthropods are classified by biologists according to the number of legs, antennae, and body regions they have. There are therefore five main groupings: crustaceans, arachnids, insects, centipedes, and millipedes. A crustacean has compound eyes, several pairs of legs (four or more), and two pairs of antennae, with a body divided into two main parts. There are about 32,000 species of crustaceans (lobsters, crayfish, crabs) who get their name from the hard case or "crust" they wear. Crustaceans eat other invertebrates, almost always live in water, and vary greatly in size. A few crustaceans, like the wood louse (also called pill bug), live on land.

Arachnids. The arachnids, which include spiders, mites, ticks, scorpions, and horseshoe crabs, have only four pairs of walking legs and no compound eyes. Their bodies have two main regions. Most live on land and feed on insects and other small animals. Most are harmless although some species are poisonous.

Insects. Insects are the most successful invertebrates on land and make up the largest class of arthropods. Well-known examples are bees, ants, grasshoppers, butterflies, moths, and the housefly. Insects live in nearly every habitat on Earth and total at least 800,000 species. All insects have a segmented body that is divided into three regions (head, thorax, and abdomen). The head contains the mouthparts and sense organs (often compound eyes); the thorax (the part of an arthropod's body where the legs are attached) has three pairs of legs or one or two pairs of wings.

Centipedes and Millipedes. The last two arthropod groups, centipedes and millipedes, both have cylindrical segmented bodies with many joined legs and antennae. The main difference between the two is that centipedes are poisonous and have one pair of legs attached to each segment. Millipedes have two pairs on each body segment. Both live in dark, damp places, but centipedes capture and eat other invertebrates, while millipedes feed mainly on decaying plant material.

BENEFICIAL AND HARMFUL

Arthropods are an extremely diverse group of invertebrates. They crawl, swim, run, and fly. They produce honey, silk, and other valuable products and provide the main meal for many of the fish that humans eat. They pollinate flowers and crops and recycle soil nutrients. Although beneficial, they can also be harmful. They can cause illness and even death to humans with their poison, and they can destroy our crops. Overall, however, arthropods are an integral part of many ecosystems (an area in which living things interact with each other and the environment), most of which would collapse without them.

[*See also* **Arachnids; Crustaceans; Insects**]

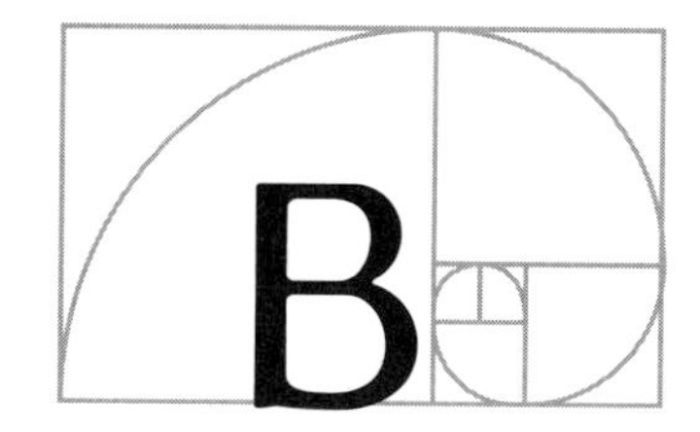

Bacteria

Bacteria are a group of one-celled organisms so small they can only be seen with a microscope. As the most abundant form of life on Earth, they live in every habitat—in the air, soil, and water, as well as within the bodies of living things. As the simplest of all organisms, they reproduce asexually by splitting in two. Although bacteria are often associated with disease, they perform an enormously useful function in the natural world since they are responsible for much of the decay of organic material.

Bacteria were first discovered in 1683 by the Dutch microscopist, Anton van Leeuwenhoek (1632–1723), who called them "little animalcules." However, it was not until the middle of the nineteenth century that biologists began to understand bacteria better. Although they were considered to be animals and then plants, bacteria eventually came to be placed in the Monera kingdom since they do not have a distinct nucleus (a cell's control center). Members of the four other kingdoms (Protista, Fungi, Plantae, and Animalia) all are eukaryotic, meaning their cells have nuclei kept within a membrane. Most bacteria are single-celled and can be grouped according to three definite shapes. Rod-shaped bacteria are called "bacilli." One species of bacillus causes the cattle disease anthrax. Spherical or round bacteria are known as "cocci." Certain "cocci" can cause staph or strep infections. Bacteria that have a helical or coiled shape resembling corkscrews are known as "spirilla." Spirilla are often carried by rats and can cause a form of rat-bite fever. Bacteria can also sometimes be in the shape of a fat comma, and these are called "vibroids." A particularly nasty form of this bacteria causes cholera.

ROBERT KOCH

German bacteriologist (one who specializes in the study of bacteria) Robert Koch (1843–1910) was the first to prove that a specific bacteria (a group of one-celled organisms so small they can only be seen with a microscope) can cause a given disease, thus helping to establish the germ theory of disease. As a pioneer in bacteria and cell culture, he also established the rules, called Koch's postulates, upon which modern bacteriology is built. He won the 1905 Nobel Prize for Physiology and Medicine for his work on tuberculosis .

Robert Koch was born near Hanover, Germany, and was one of thirteen children. Although his father wanted him to be a shoemaker, Koch was able to go to school and eventually received a medical degree in 1866. After serving his country as a surgeon in the Franco-Prussian War, he settled down as a country doctor in Wollstein in 1872. It was there that he began to study anthrax, a cattle disease, since an epidemic had struck the area. No one knew what caused this disease, and Koch set himself the task of finding out. What he would do however, would be not only to teach the world about anthrax, but help bacteriology to develop as a science as well.

Koch first obtained the anthrax bacterium from the spleen of infected cattle and gave the disease to a mouse. He then transferred the infection from that mouse to another, and then to another, until he was sure that he could identify the particular rod-shaped bacterium (called a bacillus) that caused anthrax with absolute certainty. In order to do this, Koch had to essentially invent the techniques of studying microorganisms (organisms that can only be seen with a microscope). For example, he developed a technique in which he spread a liquid gelatin on glass slide plates, which enabled him to examine a pure culture and even to photograph it. What Koch also learned to do was how to cultivate bacteria, or allow it to grow and multiply, outside the living body. Koch used blood serum at body temperature

PASTEURIZATION KILLS BACTERIA

Most scientists believe that bacteria were the earliest forms of life on Earth. Today, there are more than 10,000 species that have been identified. Bacteria can live in many different environments—even under extreme conditions such as lack of oxygen. Some live in or on other organisms, and many that grow inside the bodies of animals cause disease and are known commonly as "germs." Examples of some of the more common infectious diseases caused by bacteria are strep throat, tuberculosis, typhoid fever, and tooth decay. Certain bacteria can also cause milk and wine to go bad. In the nineteenth century, the French chemist Louis Pasteur (1822–1895) discov-

to do this, and was therefore also able to trace the entire life cycle of the anthrax bacterium. The specific techniques that Koch pioneered would serve as a model for others to follow.

Before Koch, many others had put forth the germ theory of disease, arguing that certain diseases are caused and transmitted by specific microorganisms. However, until Koch, no one had been able to prove this theory. With his identification of the anthrax bacillus, Koch not only demonstrated that this theory was true, but he also established the rules for properly identifying the cause of a disease. Called "Koch's postulates," these rules guided many a researcher in the right direction, and they still hold true today.

First, Koch determined that the suspected microorganism must be found in the infected animal. Second, after being cultured, or grown, it must be able to reinfect a healthy animal with the same disease. Third, the exact same microorganism must be found in the second animal that infected the first. Using his own techniques and rules, Koch isolated the cause of cholera epidemics in Egypt and India and discovered the tubercle bacillus that causes the lung disease, tuberculosis.

Until Koch isolated this microorganism, science was baffled by tuberculosis, not knowing how or whether it actually spread or whether it was simply hereditary. Once he identified the bacillus, however, he was able to show that tuberculosis was caused by a germ that could be carried in the air and passed from one person to another. Until Koch had proven the germ theory of disease once and for all, the science of medicine could not really progress much beyond what it was centuries before when people believed in spontaneous generation (the idea that living things can come from nonliving matter, such as maggots from rotting meat). After Koch, the science of microbiology, which is the study of things that can only be seen with a microscope, could really begin to make progress.

ered that heat would kill bacteria. When he examined good wine, he noticed that it contained yeast cells that caused the process of fermentation and therefore produced alcohol. However, when he looked at sour wine under a microscope, he saw bacteria as well. His successful heat remedy to kill unwanted bacteria has come to be known as pasteurization. It was Pasteur who also discovered the bacterial origin of certain diseases like anthrax.

THE BENEFITS OF BACTERIA

Many bacteria are harmless and essential to the well-being of certain ecosystems (an area in which living things interact with each other on the environment). Bacteria have always played a vital role in breaking down organic material after they are dead, and because of bacteria, nature is able to recycle that material's basic chemicals back into the environment. It is bacteria that immediately break down an animal's waste products. Without bacteria, the leaves that fall to the ground or the grass that is cut would never rot and get reused by nature. Just as important, the environment would be cluttered and smelly without bacteria decomposing organic matter. Bacteria even help animals digest the food they eat.

Humans have mastered killing unwanted, disease-causing bacteria (usually with antibiotics) and have learned to use bacteria in many helpful ways. Some bacteria produce desirable chemicals such as ethyl alcohol and acetic acid, while others are used in the production of food products such as cheese, butter, coffee, wine, and cocoa. Bacteria are used in the manufacture of silk, cotton, and rubber, and help produce useful medical substances like insulin, antibiotics, and interferon. Certain bacteria help clean up oil spills in the ocean by breaking up the oil into its harmless components. Bacteria have little trouble reproducing, since each bacterium can do so on its own. A bacterium reproduces by a process called "binary fission," meaning that it splits in two, making a pair of identical cells. Under proper conditions, bacteria do this about every twenty minutes.

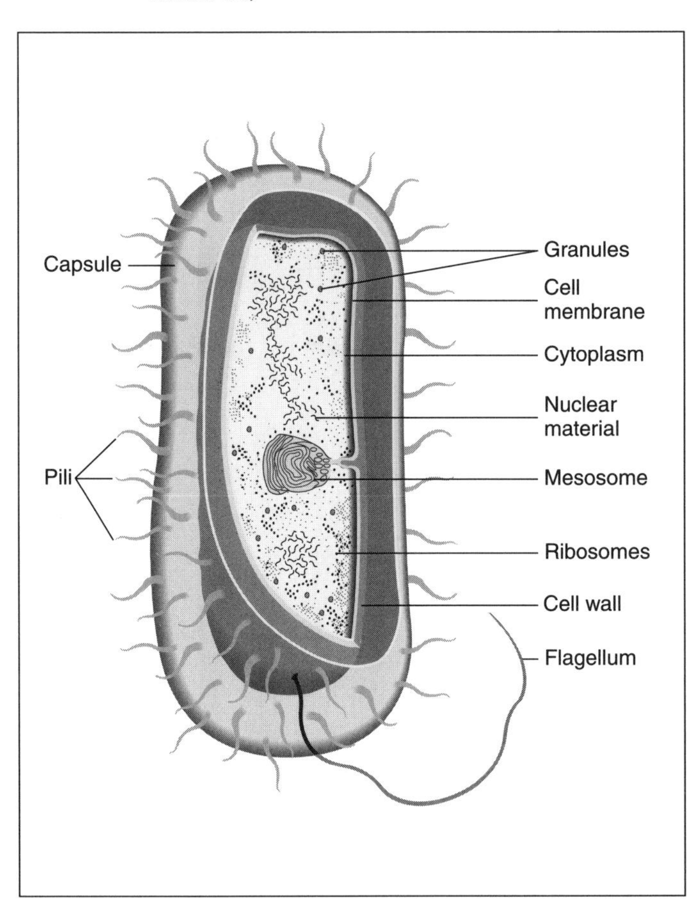

An illustration of the anatomy of a typical bacterium. (Illustration by Hans & Cassidy. Courtesy of Gale Research.)

While certain bacteria are responsible for human misery and even death, others are absolutely essential to the mechanisms of the natural world. Therefore, much of the proper functioning of nature is attributable to vastly abundant but virtually invisible organisms called bacteria.

[*See also* **Antibiotics; Fermentation**]

Biodiversity

Biodiversity, or biological diversity, is a broad term that includes all forms of life and the ecological systems in which they live. Biodiversity also refers to the degree of variety found in life and ecosystems (an area in which living things interact with each other and the environment). It suggests a level of healthy or balanced diversity of habitat, as well as animal and plant life. Biodiversity has also been described as the total richness of biological variety in a given place and can serve as a measure of the number of different types of organisms that live there. The opposite of biodiversity is the extinction of a species.

EDWARD WILSON COINS THE TERM BIODIVERSITY

As first put forth by the American biologist, Edward O. Wilson (1929–), biodiversity has become a standard against which scientists can measure the negative impact or damage that modern human societies have inflicted upon Earth and its creatures. The growing threat to biodiversity posed by human activities makes it one of the most important aspects of the present global environmental crisis that some scientists say we are experiencing. Humans have already caused permanent losses of biodiversity through extinction of certain species. Often this is caused indirectly by destroying a particularly distinctive natural environment in which these species lived. Some ecologists predict that unless there are major changes in the ways that humans affect ecosystems, there will be much larger losses of biodiversity in the near future. In the past, humans have hunted some creatures like the dodo bird into extinction, and sometimes simply crowded others out of their natural habitats. It is known that some 200 species of plants native to the United States already have become extinct, and that as much as 20 percent of all bird species around the world have gone extinct. Wilson estimates that at least 27,000 species of plants and animals are becoming extinct every year by cutting down the rain forests in Brazil's Amazon region. Since extinction is the permanent death of a species, the world will lose an enormous amount of biodiversity as the years pass.

Edward O. Wilson brought the world's attention to biodiversity and how human actions are causing its losses through extinction. (Reproduced by permission of Edward O. Wilson.)

EDWARD OSBORNE WILSON

Founder of the modern field of sociobiology (the study of what biologically determines social behavior) and the only person to receive both this nation's highest scientific and literary awards, American entomologist Edward O. Wilson (1929–)is one of the great naturalists in American history. His work has influenced the field of animal taxonomy (the science of classifying living things), and the discovery of pheromones (chemicals that are released by an animal and which affect another animal). He also has become a powerful voice condemning the steady loss of biodiversity (a broad term that includes all forms of life and the ecological systems in which they live).

Edward O. Wilson was born in Birmingham, Alabama, and is a descendant of farmers and shipowners. Always drawn to the outdoors, he decided at the age of seven to become a naturalist explorer. His plans changed, however, when he suffered an injury to his right eye while fishing, which made him change his plans. By the age of ten, after reading a *National Geographic* article titled, "Stalking Ants, Savage and Civilized," Wilson decided to become an entomologist. (Entomology is a branch of zoology that deals with insects.) After studying biology at the University of Alabama, he received his bachelor's degree in 1949. By the age of twenty-six, he had earned a Ph.D. in biology from Harvard University, where he continues to work. His early work led taxonomists (classifiers of organisms) to revise their procedures, and by the late 1950s, Wilson had begun to study how social insects, like ants, communicate. His work with the venom glands of fire ants led to his discovery that animal chemicals called pheromones are used to communicate complex instructions. He later wrote that pheromones were "not just a guidepost, but the entire message." In the 1960s, he offered the theory of species equilibrium, which demonstrates that the number of species on a small island would always remain constant.

To biologists, biodiversity is not just a blanket term for the natural biological wealth found on Earth or a description of a healthy and richly varied ecosystem, but it is something that is good biologically. In other words, biodiversity is something to value, to maintain, and to try to protect. Both the term "biodiversity" and the concept were born in 1986 during the National Forum on Biodiversity, held in Washington, D.C. and sponsored by the National Academy of Sciences and the Smithsonian Institution. Most attribute both the word and the concept to Wilson. In his extensive writings, Wilson argues that the notion of biodiversity has a value of its own whether or not it is related to humans in any way. This argument brings up the ethical question of whether mankind has the "right" to do whatever it wants with Earth and its creatures—even to the

In the 1970s, Wilson's work on the biological basis of social behavior caused a major controversy. In his 1975 book, *Sociobiology: The New Synthesis,* Wilson suggested that the way people act or behave in certain situations can sometimes be accounted for by their genetic makeup. This daring idea caused a great stir among many colleagues who argued that similar ideas had led to notions that people were "biologically determined" to act a certain way and were therefore not responsible. Wilson's critics also said that such an idea could lead to some of the policies followed by Nazi Germany during World War II (1939–45) when it claimed that certain races were genetically "inferior" to others. In 1978, Wilson defended his theory with his book *On Human Nature* in which he more fully explained his views. In this book and others, Wilson made a good case for not ignoring the role of biology in trying to understand human behavior.

In the 1990s, Wilson also became closely identified with the notion of biodiversity. Technically, this word means the variety of life or the biological diversity of species in an area. However, biodiversity also has come to mean the richness and balance of an ecological system. It does not necessarily mean huge numbers of life forms, but rather suggests equal numbers of individuals of different species. Biodiversity not only has commercial value to people but, argues Wilson, its continued loss is a signal of the worsening health of the environment. Wilson is very concerned with trying to minimize what he sees as mass extinctions that are happening in the modern world. As huge areas of rain forest are destroyed each year, species are going extinct in record numbers as they lose the habitat that support them. As people continue to drive plant and animal species to extinction, they also are killing the spirit and the beauty of the natural world. Today, Wilson is doing his best to prevent this from happening.

point of destroying them. On a more practical level, Wilson says that preserving biodiversity has very specific benefits to humans, as illustrated by the fact that in the United States, one-fourth of our prescription drugs have active ingredients obtained from plants. Wilson gives two dramatic examples. The first is the rosy periwinkle plant that became the source of two of the most effective anticancer agents ever discovered. The other is cyclosporin, an obscure fungus that lives only in Norway. This fungus became a powerful immunosuppressant drug and is entirely responsible for our ability to do organ transplants. More generally, a biologically diverse planet is simply more liveable and functional since nature, if properly balanced, keeps water and air cleaner, recycles nutrients, removes waste, and even controls erosion.

Biodiversity is a very broad term. Overall, there are three general kinds of biodiversity: ecosystem or habitat diversity, species diversity, and genetic diversity. The diversity of an ecosystem refers to the variety of actual ecosystems or places where organisms live in a certain region. Thus a region with forests, streams, ponds, grasslands, and farmland is more biologically diverse than one that has only open agricultural fields. It goes without saying that when an ecosystem or habitat disappears, a great number of species disappear as well. Species diversity refers to the ability of many different types of organisms to exist at a certain time. Finally, genetic diversity refers to the level of variability that is found among individual members of a single species. Variability, or variation, refers to the natural differences between living things, and it is a very important phenomenon since it allows a species to adapt to environmental changes over time through evolution.

The threat to our planet's biodiversity is different from most other ecological problems in that it is irreversible. Once a certain species has gone extinct it has disappeared forever and its entire heritage that took millions of years to evolve is lost. Since scientists admit that the vast majority of species have not yet even been identified, humans may be destroying species about which nothing in known, and therefore never will. Finally, by trying to solve some of the problems associated with threats to biodiversity, people gain new knowledge about managing the Earth and using its natural products for the benefit of all.

Biological Community

A biological community is a collection of all the different living things found in the same geographic area. The community may be small or large, but it always consists of different types of living things interacting with one another in the same particular area or habitat. A biological community is held together by the relationships among its members.

Biologists can study the natural world from many different ecological perspectives. That is, they can study life or living things according to the size and type of certain groupings in which life is found. At the species level, biologists study the same type of organisms that are capable of mating and producing offspring together. For example, one species of bird is a cardinal, another species is a blue jay. They do not breed with each other so they are different species. One level up from species is a population. A population is made up of all the members of a species who live in a given area at the same time. All the different populations living and interacting within a particular geographic area make up a biological (or biotic) community. To keep going up the scale, the living organisms in a community, together with their nonliving (or abiotic) environment, make up an ecosystem. At this point, in theory, an ecosystem (and the biological community that makes up its living component) can be as small as a few mosquito larvae living in a rain puddle or as large as a prairie stretching hundreds of miles. A very large ecosystem is called a biome.

A biological community is made up of all the populations living in an area. As a community, these populations live together and interact with one another in many ways. In fact, relationships are what biological communities are all about. Some populations eat food that is produced by another. Others get their homes from other populations. Competition is also a way of interacting, and populations may compete with one another for food and shelter. Biological communities are thus tied together by a food chain or food web (consisting of who eats what).

When studying a biological community of any size, biologists use certain categories to describe these communities. Communities may have populations that interact in antagonistic ways, meaning that the relationship is detrimental or harmful to one or both species. They may have commensal relationships, meaning one species benefits while the other is unaffected. Or they may have mutualistic or cooperative relationships in which both species benefit. Not surprisingly, in the real world of biological communities, these relationships sometimes shift.

Biologists also take note of productivity and trophic levels in a biological community. Productivity describes the amount of biomass (or living matter) produced by green plants as they capture sunlight and create new organic compounds. A tropical rain forest will have a very high rate of productivity compared to that of a desert. Trophic levels describe the level or position of a given species on the food chain. Since a hawk eats mice, which eats plants, the hawk is at a higher trophic level than the mouse. Interestingly, the same species can occupy different trophic levels in different food chains found in different biological communities. Therefore in a food chain containing a killer whale, a leopard seal, and an emperor penguin, the leopard seal is a tertiary consumer, or at the third trophic level, since it eats the penguin, which eats fish, which eats krill. However, in a chain consisting of a killer whale, a leopard seal, and an adelie penguin, the seal is at the second level since it eats the adelie penguin, which is a primary consumer because it only feeds on krill.

The technical term "abundance" refers to the total number of organisms in a biological community, while the expression "diversity" is a measure of the number of different species in a given community. "Complexity" is the term used to describe the variety or number of different ecological niches in the community. A niche is the role a living thing plays in its environment. Niche also refers to the place an organism can fit into, or the way it makes its living. Overall, the concept of a biological community is a practical and very useful scientific way to organize, categorize, and learn more about the dynamic goings-on in a particular habitat.

[*See also* **Abiotic/Biotic Environment; Food Web/Food Chain; Niche**]

Biology

Biology is the study of life or living things. Biology includes a huge range of subjects, all of which are based on studying the way living things work and interact with everything around them.

Since biology is concerned solely with living things, it is important to know what the characteristics of being alive are. All living things show four main characteristics. First, they have metabolic processes, which means they conduct some sort of chemical reactions to take in nutrition, process it, and eliminate waste. Second, they have generative processes, which means they are able to grow and to reproduce. Third, they have responsive processes, which means they react to stimuli and can adapt to changing conditions. Fourth, they have control processes, which means they can coordinate their metabolic processes in the right order and can regulate them as well. If something demonstrates every one of these characteristics, we can say that it is alive or that it is an organism. An organism is, therefore, any single living thing that demonstrates the characteristics of life.

Organisms or living things can be studied from many different levels or aspects. At the molecular level (a molecule being a chemical unit made up of two or more atoms linked together), biologists study the complex of chemicals that work together in living things. A molecular biologist would study such molecules as proteins and nucleic acids and try to discover exactly how they work in a living thing. In terms of how living things are constructed and function, all living things are made up of cells. Since the cell is the basic unit or building block of all living things, cell biology is the next level of biological study. Some organisms are made up of a single cell, while others are composed of trillions. The next structural and functional level is that of tissues and organs, and beyond that is the complete organism itself. At this point, biologists often focus on one particular group of living thing, such as plants (botany), animals (zoology), or a certain type of animal, like insects (entomology). Just as all living things have the same characteristics listed above, so all are governed by the same, few biological principles. One of these is the notion of homeostasis. The word "homeostasis" means "staying the same," and all living things need to stay the same or maintain a constant internal environment. This idea was first suggested by the French physiologist, Claude Bernard (1813–1878), who showed that an organism has control systems that enable it to keep its metabolism (internal chemistry) within certain limits, especially when things around it are always changing. Another biological principle is that all living things are made of the same materials, share the same functions, and have a common origin. Moreover, all follow the laws of heredity and possess genes that are the basic unit of inheritance. This leads to the principle of evolution by natural selection, which explains how life evolves and becomes different over time. The fact that life is very different despite its common origins leads to another principle—diversity.

Altogether, biology or the study of life can examine its subject from a highly focused and specialized point of view or from the larger viewpoint of what all living things have in common. At the same time, it can be highly theoretical (such as plant taxonomy, which is the science of classifying or naming plants), or it can be very practical (such as plant breeding or wildlife management). In the future, biology will have to cope with and try to solve some of the more important twenty-first-century issues. These involve problems related to increasing human populations and the need for increased food production. Biology is also the foundation of all medical advances, and this century will most likely focus on the genetic aspects of diseases. Finally, biology will have to face the pressing ecological problems that a growing, highly mechanized world creates. As a result of these issues, many scientists feel that the twenty-first century will necessarily be the biological century.

[*See also* **Botany; Ecology; Evolution; Genetics; Physiology**]

Biome

A biome is a large geographical area characterized by a distinct climate and soil as well as particular kinds of plants and animals. Biologists have divided the globe into six main biomes—which they also call major life zones—and have named them after the dominant type of vegetation that grows there. These six terrestrial (or land-based) biomes are the rain forest, tundra, taiga, temperate deciduous forest, desert, and grasslands. There are also three other biomes for the aquatic environment: freshwater, saltwater, and estuaries.

Biomes can be described as environments that have a lot of things in common, such as climate, vegetation, and animal life. Grouping the world's environments into separate biomes helps biologists to better understand them and to organize what has been discovered about them. Each biome is home to certain groups or communities of plants and animals that have adapted to its particular temperature and precipitation (amount of rainfall). Since a biome is necessarily a large geographic area, it usually supports a wide variety of life within the same major zone. A biome can be made up of different habitats or places where organisms live; altogether, the biomes of Earth make up the biosphere, which is the region of Earth that support life.

RAIN FOREST BIOMES

Among the six major land biomes, the biome known as the rain forest supports more species than any other, containing about half of the world's plant and animal species. Rain forests are commonly found in tropical areas close to the equator in areas of very heavy rainfall and constantly warm temperatures. They are considered one of the most biologically diverse ecosystems (an area in which living things interact with each other and the environment) on Earth. In a typical rain forest, almost every plant has another plant growing on it and the tallest trees create a canopy or rooflike effect. Actually, there are four other separate layers or levels or vegetation beneath this canopy (including lianas) that can be as thin as a rope or as thick as a tree trunk. Other plants called epiphytes (like ferns and bromeliads) grow in clumps on tree branches, and at ground level there is little vegetation since little sunlight can penetrate to the forest floor. High temperatures, constant moisture, and high humidity cause anything dead to rot very quickly. There are few ground-dwelling animals other than ants and termites; and most birds, mammals, reptiles, and amphibians live somewhere in the multilevel canopy. Fifty-seven percent of all tropical rain forests are located in South America where they are steadily being destroyed. Increasing population pressure has resulted in vast expanses of these forests being cleared for farming, only to prove unsuitable for agriculture. Constant rains quickly wash away the soil's nutrients (which were contained in a thin layer of humus), and heavy fertilizers must be added regularly. When the habitat is destroyed in this fashion, the variety of life that used to thrive in these areas disappears as well.

TUNDRA BIOMES

No environment on Earth is more different from a rain forest than the biome known as tundra. It is the coldest of all terrestrial biomes. Characterized by extremely long, severe winters and short, cool summers, tundra has been described as a polar desert. What makes it unique is the existence of "permafrost"—a condition in which the subsoil, or Earth below the top few feet of soil, remains permanently frozen. The small amount of precipitation this area does receive is trapped at the surface by the permafrost below and forms bogs in the summer. Since the subsoil never thaws, no trees or shrubs can survive, and the vegetation that manages to live there is dominated by mosses, lichens, and other plants that do not grow very large and which can reproduce very quickly during the short growing season. Few animals live on the tundra, although the Snowy owl and Arctic hare are exceptions. There are, however, vast swarms of flies and mosquitoes that appear when the few herbs and grasses flower and attract migratory birds. Although the tundra is a relatively simple environment, it is easily damaged and very slow to recover. The northern parts of Canada are composed primarily of tundra.

TAIGA BIOMES

Taiga, also called the coniferous or evergreen forest biome, supports a diverse and complex community of organisms. Although its winters can be very cold, its summers are longer than those of the tundra, allowing its soil to thaw completely. The taiga is primarily an environment where evergreen trees flourish since they are specially adapted to survive the long, cold winters. The evergreens' needle-shaped leaves prevent water loss during the harsh winter months and their flexible branches allow heavy snow to slide off without breaking. The ground below them has little vegetation and is home to mice, squirrels, porcupines, grouse, warbler birds, wolves, and moose. This biome is particularly common in southern Canada and is particularly susceptible to the effects of acid rain.

TEMPERATE DECIDUOUS FOREST BIOMES

The temperate deciduous forest biome has what we consider to be normal, well-defined seasons and is named after its most common feature—deciduous trees or trees that drop their leaves in the autumn (like oak, maple, hickory, and ash). Most of the eastern part of the United States is typical of this biome, and it is characterized by a moderate climate and fairly high rainfall. Although it is a single biome, temperate deciduous forests can be quite different according to their specific location. All support a wide variety of animal life. Insects are often abundant and some birds live there year-round, despite its winters. In this biome, leaves that fall from deciduous trees decay and result in a deep, rich humus layer that provides nutrients and conserves water. While many different mammals were once present in great numbers in this biome, today they have been reduced substantially by humans. Still, there are many raccoons, deer, moose, bobcats, a large variety of birds, and invertebrates.

DESERT BIOMES

In those parts of the world where rainfall is very low and irregular, a desert biome can be found. A desert is not only very dry, with a total rainfall between 1 to 10 inches (2.54 to 25.38 centimeters) a year, but it is also usually very hot. While most would guess that the average temperature of a desert is extremely high, it actually fluctuates or changes dramatically. At night the land cools rapidly because there is seldom any cloud cover to insulate and keep the desert heat from radiating off into space. In a desert like the Kalahari in Africa, this fluctuation can go from well over 100 °F (37.78 °C) to just above freezing. That is why a desert biome is characterized primarily by its low rainfall and not its temperature. Plants that grow in a desert are highly specialized and able to gather and store water in many different ways. For example, the leaves on a cactus became spines, while other cactus have extremely deep and spreading root systems. Despite this harsh environment, many other plants and animals live in deserts, although the desert reptiles, small mammals, and spiders and scorpions usually spend the day in burrows or shaded areas and become active at night (meaning they are nocturnal). Some animals, like the camel, have also adapted to the lack of moisture in interesting ways. Contrary to what most people believe however, the camel does not store water in its hump. Rather, it stores energy in the form of fat, and by concentrating all of its fat in one part of its body, it is allows its body heat to escape more easily from the rest of its body. Camels also can allow their body temperature to fluctuate more when water is scarce.

GRASSLAND BIOMES

Certain parts of the world are too rainy for a desert but not wet enough to support a real forest. These areas are called a grassland biome. Grasslands (also called savannahs) have hot summers and cold winters and receive between 15 and 30 inches (38.1 and 76.2 centimeters) of rain a year. Abundant grasses replace trees (that seldom grow in these fertile places). If the soil is deep and rich, tallgrass grows, while shortgrass grows in thinner soil. In the United States, grasslands are called prairies; in South America they are the pampas; in Asia they are called steppes; and in South Africa they are known as veldt. Whatever the actual name, these flatlands are covered by fast-growing grasses that are enormously efficient at converting sunlight into plant tissue. Animals common to grasslands are the prairie dog, coyote, grasshopper, rattlesnake, pronghorn antelope, and meadowlark. Grasslands often catch fire during dry times, but the roots usually remain alive allowing grass to regrow there. Grasslands are amazingly fertile and can support regular farming if treated properly, but they also turn into dustbowls with wrong management.

FRESHWATER, SALTWATER, AND ESTUARY BIOMES

Earth's water systems have their own types of biomes divided simply between freshwater, saltwater, and estuary biomes. Freshwater biomes either move (streams and rivers) or stand (lakes and ponds), and the types of fish and other organisms that live in them are determined by temperature and other variables like oxygen and minerals. The marine, or saltwater, biome is the largest of all the world's biomes, covering nearly two-thirds of the planet's surface. With a salt content of between 3.0 and 3.7 percent, saltwater biomes support fish and other organisms that have adapted to this specialized environment. An estuary is a partially closed-off part of the sea that is fed by fresh water. These include bays and tidal marsh inlets. Since the salt content varies greatly, relatively few species can tolerate estuaries but those that do are extremely productive. Because estuaries are fairly shallow, sunlight can penetrate them completely making them ideal habitats for algae, grasses, oysters, and barnacles.

Human influence has altered the world's biomes. Some biomes no longer have the diversity of life that they once contained. The quantity and diversity of mammals found in North America today can in no way compare to those that existed there some five hundred years ago. Nearly all of the original forest trees are also gone. Human activity is thought to result in a global temperature rise, known as the greenhouse effect. This rise in temperature could also alter the nature of each biome causing the extinction of plants and animals.

[*See also* **Chaparral; Desert; Ecosystem; Forests; Grasslands; Ocean; Rain Forest; Taiga; Tundra; Wetlands**]

Biosphere

The biosphere is that part of Earth that contains life. This worldwide ecosystem (a area in which living things interact with each other and the environment) is made up of the land, water, and atmosphere that support life, and includes every part of Earth where life exists. Its many parts are linked together by nutrient cycles.

The biosphere can be described as all of the world's ecosystems, or the worldwide ecosystem. As such, it consists of Earth's lithosphere, atmosphere, and hydrosphere. The lithosphere includes all of Earth's surface layers of solid substances like soil and rocks. The atmosphere is the envelope of gases or air that completely surround the planet. The hydrosphere includes all of the lakes, rivers, and oceans on its surface. All of the elements and forces in these three spheres are constantly part of the many chemical, biological, and physical processes that make the entire "bio-sphere" or "life-sphere" what it is. To date, no other planet besides Earth has been discovered that contains a biosphere or living world.

EDUARD SUESS COINS THE TERM BIOSPHERE

The term biosphere was first used by the Austrian geologist Eduard Suess (1831–1914) in 1875 to describe that part of the Earth that contains life. This concept had little impact on the scientific community until it was discussed by the Russian mineralogist, Vladimir I. Vernadsky (1863–1945) in his 1926 book. Vernadsky argued that in order to study the biosphere, scientists from many fields like geology, chemistry, and biology had to work together.

Earth's biosphere can be considered to be thousands of feet thick—from the bottom of the oceans to about 30,000 feet (9.144 kilometers) above sea level. However, given the size of the entire Earth, this is a fairly thin layer. In fact, most of the life that does exist within the biosphere can be found in the narrow band just below sea level to about 20,000 feet (6.096 kilometers) up. That height is about the limit for animals and most plants to live. Scientists consider Earth's biosphere a closed system in which the only thing ever added is sunlight. Living things in the biosphere need energy and nutrients and an appropriate environment in order to live and reproduce. The Sun is the ultimate energy source for the biosphere, and its light energy is captured by green plants that use the process of photosynthesis to change light energy into chemical energy. This energy then passes through the biosphere from plants to other organisms and then to others again. It is eventually either lost in the form of heat to the environment or stored for a long time in organic molecules, such as carbon atoms are locked in coal underground. The biosphere can, therefore, be seen as a group of tightly interconnected recycling systems and subsystems that affect and influence each other.

THE BIOSPHERE CONSIDERED AS A GLOBAL CONCEPT

The global concept of the biosphere has become increasingly useful and important as technology allows the world to be considered in a truly global way. Spacecraft can not only provide images of the entire globe, but also can monitor the many systems that make up Earth's total environment. Today, therefore, the term biosphere is used most often when discussing the health of Earth. Current research focuses on the effects that human activities have on the health of the global environment (the biosphere). The pressures and demands of increasing populations have resulted in two major threats to the biosphere's well-being: the loss of natural resources and the effects of pollution. As more and more of the natural environment is destroyed, such as the continuing destruction of tropical rain forests, many scientists fear that the balance of nature may be so upset that it may permanently harm or change the biosphere. Many also fear that the pollution by-products of everyday modern life could become too great for the biosphere to bear and may permanently damage its systems. Activities such as fuel consumption (burning coal, oil, and gas) and the use of fertilizers could increase the levels of carbon dioxide, nitrogen, and phosphorous in the atmosphere and seriously alter the natural balance of the life-sustaining biosphere.

The concept of the biosphere is also at the center of a scientific debate as to whether Earth and its biosphere should be considered a living organism with self-regulating mechanisms. Called the "Gaia hypothesis," this idea was first put forth in the late 1970s by the British scientist, James Lovelock (1919–), who argued that the biosphere is able to create and maintain the environment that most favors its own stability—as long as it is left alone. Lovelock argued that by tampering with and sometimes altering Earth's environmental balancing systems, humans are placing themselves and their planet at risk. He and his followers argue that such phenomena as global warming, or the greenhouse effect (caused by too much carbon dioxide in the atmosphere), and ozone depletion from pollution (the resulting decrease in protection from harmful ultraviolet radiation from the Sun, are evidence of a growing risk. Whether the Gaia hypothesis is right or wrong, it has been very useful in generating concern and research about people's influence on the worldwide ecosystem known as the biosphere.

[*See also* **Photosynthesis; Respiration**]

Birds

A bird is a warm-blooded vertebrate (an animal with a backbone) that has feathers, a beak, and two wings. Its most unique feature is the ability to fly, although not every bird is able to fly. All birds hatch from eggs and

JOHN JAMES AUDUBON

American ornithologist (a person specializing in the study of birds) John James Audubon (1785–1851) was the most famous artist and naturalist in nineteenth-century America. He not only kept finely detailed studies of birds, but produced the first modern atlas of ornithology. This atlas is also considered to be one of the most beautiful natural history books ever made.

John James Audubon was born in what is now Haiti, the son of a French sea captain who owned a plantation on that island. After spending his first few years there, he moved to Paris, France, and studied painting for a short time. Audubon was eventually sent to America when he was eighteen years old in order to avoid being inducted into the army of Napoleon. Throughout his boyhood, he continuously collected and sketched birds, plants, and insects. He also developed the habit of keeping detailed notes of whatever he observed. At his father's estate at Mill Grove, near Philadelphia, Pennsylvania, he conducted what appear to be the first known bird-banding experiments in North America. In order to learn more about the movements and habits of a bird called the Eastern Phoebe, he caught them, tied bits of colored string on their legs, and was able to prove that they returned to the same nesting sites the following year.

After Audubon married Lucy Blakewell in 1808, he tried to become a storekeeper, but found himself unable to stay indoors long enough to run a store. As a natural-born outdoorsman, he was unable to make himself stay away from studying the things he loved. After his business failed, and he went bankrupt (and even spent some time in jail), he decided to pursue what he really loved and dedicated himself to what he now called his "great idea." He would travel throughout America and draw, in life-size scale, every bird on the continent. Thus, at the age of thirty-five, Audubon set off, with his wife's blessing, to pursue his ornithological dream. Audubon's wife worked as a governess and teacher to help support him, although he would sometimes paint portraits and street signs when they needed money. For five years Audubon traveled the American wilderness painting its bird life. Where most bird painters before him had worked from long-dead, stuffed

their bodies have evolved a wide range of adaptations that enable them to fly. Birds are found in nearly all parts of the world.

CHARACTERISTICS OF BIRDS

From the soaring eagle to the flightless penguin, from the clumsy loon to the graceful swan, birds come in the widest variety imaginable. All have several characteristic features that distinguish them from other ani-

specimens and drew them in a not very true-to-life way, Audubon pioneered the use of fresh models. He would shoot a bird and wire into a life-like pose in order to try and capture it accurately. It must be remembered that in the early nineteenth century, little thought was ever given to the conservation of birds or any other animals, and it was common for Audubon to shoot several of the same species until he found one that he considered perfect for painting. From the beginning, Audubon's "great idea" ambitiously included the goal of painting every bird in its real, life-size dimensions. He was able to do this for most birds by having a single book page measure more than 3.5 feet (1.07 meters) by 2.5 feet (0.76 meters). This was larger than any book published to that time. For the very large birds like the whooping crane, he would paint them life-size but have their heads bending toward the ground, so that the drawing would fit on a page.

By 1825, Audubon had compiled a spectacular set of bird paintings, but when he could interest no American publisher in his work, he went to England. There his work was recognized, and he began to enlist "subscribers," either individuals or institutions, who agreed to buy the book when published. The actual production of his great work took many years since the plates (a full-page book illustration) were all hand-colored. The first volume of *Birds of America* appeared in 1827, and the fourth and last volume was published in 1838. Altogether, the books contain 435 hand-colored pictures. Later, Audubon published a five-volume work of bird descriptions to accompany the illustrated work. The production of *Birds of America* was extremely expensive, and a subscriber paid approximately $1,000 for the set, an extremely high price in the early nineteenth century.

Despite the fact that Audubon probably killed more birds than most anyone of his time, he had dedicated his life to capturing their beauty and the essence of their liveliness, and he truly must have also loved birds more than anyone of his time. Today, his name is linked to that of a modern conservation organization, the National Audubon Society, and there are many local Audubon societies throughout America dedicated to learning about and conserving birds.

mals. Only birds have feathers. While other types of animals may have hair or scales or a shell covering their bodies, feathers are unique to birds. As with just about everything in the way they have evolved, a bird's overlapping feathers give it the ability to fly. The most important factor in flight, whether it is a bird or an airplane, is weight. The heavier something weighs, the more energy is required to get it off the ground and keep it in the air. Feathers are therefore, by design, a lightweight covering for a bird. Like human hair or a bull's horn, feathers are made of a protein called keratin. Although feathers vary in size, shape, and texture, they share a basic frame or structure: a base like a tube that goes up into a central shaft that itself branches into vanes. A bird actually has several different types of feathers. Short, fluffy, down feathers that are closest to the skin serve as insulation. Powder feathers produce a type of powder that makes a bird waterproof. Contour feathers give the bird a streamlined shape. Flight feathers are strong and provide lift much like an airplane wing. Tail feathers are stiff and provide steering, balance, and control. Altogether, a bird's feathers are the perfect covering for its wings. Birds are able to fly because when air passes over their wings, they receive an upward push or lift. Feathers also act as a good insulator and keep a bird warm by helping it retain its body heat.

All birds have two feet and one beak, although there can be some very dramatic differences. The shape and structure of both beaks and feet give a good indication of what a bird eats and what type of habitat it lives in. The long, tubelike beak of a hummingbird is adapted for sucking nectar from a flower, while a bird of prey like an eagle has a sharp, hooked beak for holding and tearing its meal. Finches have short beaks that are thick and strong for cracking open seeds, and a typical marsh bird has an upward-curved, spoonlike bill for sifting water. Ducks have webbed feet for swimming atop water, while hunters and meat-eaters like hawks have sharp talons that grip and kill their prey. Other birds simply have toes that enable them to perch on branches. Most birds have four toes—three in front and one in back.

Finally, all birds are endothermic or warm-blooded and hatch from eggs. Since birds are warm-blooded, meaning that they maintain a constant internal body temperature despite the temperature of their environment, they are able to live in a wide variety of climates and environments. Unlike cold-blooded animals, they do not have to slow down their activity when the temperature drops. Compared to almost any other animal, birds use up an enormous amount of energy and have a very high rate of metabolism (which could be described as the rate at which their internal motor is running). This is because flying is such a high-energy activity. Since they have a high-energy lifestyle, birds have a high-energy diet (such as seeds, worms, fruit, meat, fish) and do not eat many low-energy foods like grasses or leaves.

Birds reproduce sexually and lay eggs (one at a time) from which their young hatch. Birds don't give birth to live young since weight is so critical, and flying while carrying a developing embryo inside would probably be impossible for a female. Bird eggs contain a yolk that feeds the embryo. They are covered by a hard shell, and one or both parents must keep them warm with their own bodies for them to develop and hatch. At birth, many birds are helpless and must be fed and cared for. While they may learn certain things from the behavior of their parents, important activities like flying and migrating are instinctive.

BIRDS BODIES ADAPTED TO FLIGHT

As a bird's feathers are perfectly adapted to flight, so too is the rest of its body built to give it the two things it needs to fly: low weight and high power. Compared to almost any other animal, a bird's skeleton makes up only a small percentage of its total weight. This is because its bones are hollow, or in fact, filled with tiny air spaces. Despite this, the bird has a very strong frame since many bones that are separate in other animals are fused in birds. Other systems conserve weight as well. A bird has no separate bladder (which would fill and add weight), but rather passes its nitrogenous waste along with its intestinal waste in a single pastelike form called bird droppings. Birds get their flying power from powerful breast muscles, and are able to feed these important muscles all the oxygen they need because of their highly efficient respiratory systems (which can also store air). Birds, like mammals, also have a four-chambered heart which keeps oxygen-rich blood from the lungs separate from the blood that is returning from the rest of its body. This guarantees that the muscles get as much oxygen as they need. Birds also have an above-average nervous system since they must perform highly coordinated movements in order to fly. They also have very sharp eyesight and good hearing.

As with insects, flight provides birds with a competitive advantage. They are able to escape quickly from an earthbound predator who cannot pursue them. Birds are also able to move to another habitat if food becomes scarce. Many birds do this annually as a regular part of their life cycle in a process called migration. Some birds make this an entire way of life, migrating as much as 25,000 miles (40,225 kilometers) in one year.

THE IMPORTANCE OF BIRDS

Birds play an important role in the ecosystem (an area in which living things interact with each other and the environment). They not only provide people with food (like chicken), but they consume an enormous quantity of insects and play a major role in pollination (the transfer of male pollen to the female part of a flower). Finally, birds make sounds that people find pleasant. Bird sounds are both calls and song, with the former used to communicate with others, and the latter used mainly to attract a mate.

Blood

Blood is a complex liquid that circulates throughout an animal's body and keeps the body's cells alive. Blood transports oxygen and nutrients, carries away waste, and helps fight germs that invade the body. Each different function of the blood is carried out by a different type of blood cell.

The importance of blood to life was recognized by primitive humans, but they had no understanding of why it was so. They did not know that the cells in an animal's body are very specialized and that these cells cannot get their own food or dispose of their waste. Primitive humans also did not know that it is the blood that carries oxygen and nutrients to the cells, while also carrying away waste products, and that blood plays a major role in fighting disease by defending the body from invasions of microorganisms and parasites. These ancient humans believed that blood had mystical properties, and in many ways they were right since blood has the ability to sustain life.

SCIENCE LEARNS TO UNDERSTAND BLOOD

The first real scientific contribution that concerned blood was made by the English physician, William Harvey (1578–1657). Harvey made the major discovery that blood circulates constantly throughout the body in a one-way direction. Harvey said the heart was a pump which pushed the blood through a one-way, closed-loop circulation, and he was able to demonstrate this by tying off arteries and veins. He proved that blood did not wash back and forth like the ocean tides, but instead kept constantly moving in the same direction. Once this mechanical aspect of blood circulation was understood, science moved toward an understanding of the composition and actual function of blood.

After the seventeenth-century invention of the microscope, the Dutch naturalist Jan Swammerdam (1637–1680) discovered the red blood cell in frog's blood. In 1673 and 1674, his countryman, Anton van Leeuwenhoek (1632–1723), was the first to see and describe red blood cells in human blood. In the next century, white blood cells (known as leukocytes) were first seen, and finally platelets (tiny fragments that cause clotting) were first observed in 1842.

PARTS OF BLOOD

By the start of the twentieth century, the composition of blood was known. It was found to be composed of red blood cells, white blood cells, platelets, and plasma. Besides knowing what blood was made of, scientists also were beginning to understand each parts real function.

Red Blood Cells. Red blood cells are called erythrocytes, from the Greek *erythros,* meaning red. These cells collect oxygen from the lungs and carry it to all the cells in the body. Under a microscope they look like flattened disks, and they do not contain a nucleus (the control center for most of a cell's activities). Erythrocytes get their distinctive red color from a pigment called hemoglobin. Hemoglobin is an iron-containing protein that does the actual oxygen-carrying for the blood. People who have a diet low in iron may not have enough hemoglobin, and they would tend to tire easily since their blood does not carry enough oxygen. Hemoglobin also carries carbon dioxide away from the cells to be disposed by the lungs. Red blood cells are made in the bone marrow (a soft, fatty tissues that fills most of the bone cavities and is the source of red blood cells and many white blood cells) and live for about four months.

White Blood Cells. White blood cells are called leukocytes and form part of the body's defense against invasion. They are larger than red blood cells, and as part of the body's immune system, their job is to fight infection. They do this by engulfing and killing the invading organism.

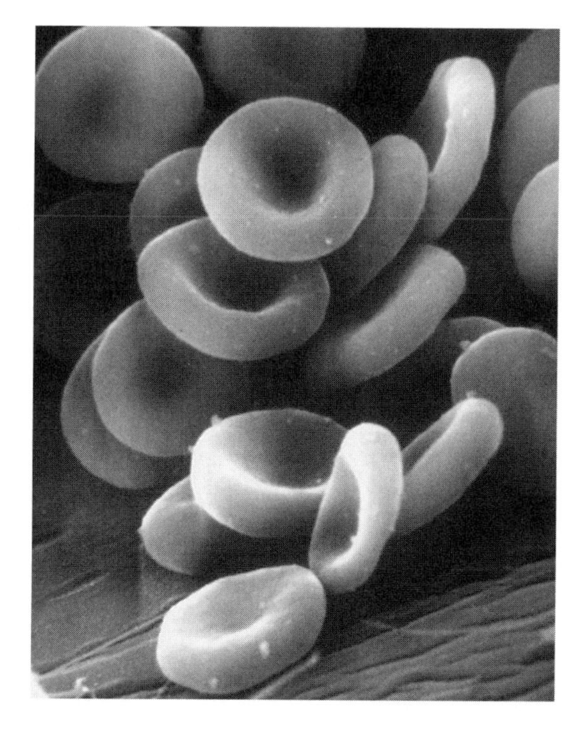

Magnified red blood cells. Anton van Leeuwenhoek was the first to see and describe these cells in the human blood. (Reproduced by permission of Phototake. Photograph by Dennis Kunkel.)

Platelets. Blood also contains tiny fragments called platelets or thrombocytes. These small fragments go into action when the body is injured. They help cause blood to clot or form a plug to stop bleeding. They also help repair damaged blood vessels. Clotting or coagulation of the blood prevents foreign organisms from getting in and minimizes blood loss.

Plasms. All of these specialized cells and cell fragments (red cells, white cells, and platelets) are suspended in a liquid called plasma. Plasma is the liquid portion of the blood. It has often been compared to seawater in the amount of sodium chloride (common salt) it contains.

BLOOD TYPES AND DISORDERS

There are slight but important differences that people have in their blood, and this accounts for what are known as blood types or blood groups. There are four major human blood groups

(A, B, AB, and O), and people can only receive blood from someone else of the same blood group. These groups are determined by certain proteins in the blood.

Blood disorders take the form of problems with red blood cells, the worst of which are hereditary diseases like sickle cell anemia. Leukemia is a type of white blood cell disorder in which these cells reproduce uncontrollably. They should only make up about one percent of the blood. Platelet disorders can result in bleeding problems or dangerous clots. Future breakthroughs in blood research may soon result in cures for these conditions, such as artificial blood or some acceptable blood substitute that will be able to carry oxygen to the cells.

[*See also* **Blood Types; Circulatory System; Heart; Respiratory System**]

Blood Types

A blood type is a certain class or group of blood that has particular properties. There are four major human blood types, which are inherited, and each of which has a characteristic protein on the surface of its red blood cells. Individuals who share the same type of proteins belong to the same blood group or type. It is essential to know a person's blood type before a blood transfusion can be given.

Since the Middle Ages (A.D. 500–1450), doctors thought that if there were only some way to replace the blood a person lost due to injury, they could possibly save lives. Once precision instruments were developed in the seventeenth century that could be used to inject one person's blood into another, blood transfusions were attempted. In far too many of these experiments, however, the results were just the opposite of what was expected. Many patients died and those that did not often became even more ill. Since no one had any idea why blood could not simply be transfused from one person to another, blood transfusions were eventually banned in most of western Europe after the late seventeenth century.

KARL LANDSTEINER DISCOVERS BLOOD GROUPS

Research on blood did continue, and in the late 1800s, several researchers noted that when blood cells from one animal or person were mixed with cells from another, they stuck together in clumps. This was called agglutination. While studying this clumping phenomenon, the Austrian physician Karl Landsteiner (1868–1943) discovered that not all blood always clumped with other blood. For example, one sample would clump with red cells from person A but not with person B. Another sample might clump red cells from person B but not from person A. Still another sample might clump them both, while yet another might clump neither. Eventually, Landsteiner was able to clearly identify four main blood groups that he named A, B, AB, and O.

Further research showed that these four blood types differed because of the type of protein that was located on the surface of the red blood cells and in the blood's plasma (the fluid part of the blood). These proteins on red blood cells came to be called antigens (a kind of chemical identification tag), while those in plasma were called antibodies (proteins that destroy foreign substances). In what came to be known as the ABO system, there are two antigens (A and B) and four blood groups (A, B, AB, and O). People with type A blood have the A antigen; people with type B blood have the B antigen; type AB people have both; and type O people have neither.

This led Landsteiner to formulate a simple pattern of who can receive what from whom. The first rule is that people in the same blood group can accept blood from each other with no ill effects. Next, blood types A and B are incompatible and cannot receive blood from each other, but they can receive blood from O (since it has no antigens). Blood type AB can accept blood from A or B (since they have both A and B antigens), as well as from O (which has no antigens). The AB blood type is therefore called the "universal recipient." Type O can give blood to all other groups, but can only receive blood from its own type. It is therefore called the "universal donor."

LANDSTEINER ALSO DISCOVERS THE RHESUS FACTOR

The Rh (rhesus) factor system is another blood group that was discovered by Landsteiner and his associates in 1940. First discovered in rhesus monkeys, it was found that about 85 percent of the human population was Rh positive, meaning that their blood cells carried the D antigen or rhesus antigen. Those that did not were Rh negative. A person's Rh factor becomes important during pregnancy. A fetus (unborn child) that is carried by an Rh-negative woman who developed Rh antibodies by previously carrying an Rh-positive baby can have its red blood cells attacked by these antibodies, resulting in death. The most common blood type in the United States is O+ (O Rh positive). It is found in 38 percent of the population. The type A+ (A Rh positive) is

KARL LANDSTEINER

Austrian-American immunologist (a person specializing in the study of the immune system) Karl Landsteiner (1868–1943) discovered the main types of human blood. His blood-typing system made blood transfusions possible and saved countless lives. Awarded the Nobel Prize for Physiology and Medicine in 1930, he also discovered the Rhesus (Rh) blood factor and that polio is caused by a virus (a disease-causing agent).

Since at least the Middle Ages (A.D. 500–1450), doctors had been intrigued by the idea that severe blood loss might be treated simply by injecting the blood of one person into another. Once instruments precise enough to be able to do this were produced in the seventeenth century, blood transfusions were attempted. Sometimes they would work and save a patient, but much more often the transfusion itself would kill the person who was receiving someone else's blood. This happened so often that blood transfusions eventually were banned in most European countries. Until the problem was taken up by Karl Landsteiner, no one knew the reason that one person's blood could not be transferred to anyone else. All anyone knew was summed up by folk wisdom which simply said that everyone's blood was different.

Karl Landsteiner was born in Vienna, Austria, and entered medical school at the age of seventeen. By the time he was twenty-three, he had received his doctorate in medicine and went to work in the field of organic chemistry, studying with some of the best chemists in Europe. By around 1896, he became interested in the nature of antibodies, which are special proteins that circulate in the blood and lock on and disable any foreign substance that enters the body. By 1900, he was studying how blood agglutinates, or clumps, together when it is brought into contact with the blood of another person. No one could properly explain why this happened, but Landsteiner believed that it was due to something unique in the blood of each individual. He then began a series of experiments that showed that there were often very different things going on when blood clumped together. For example, the blood of one person might clump the blood from

found in 34 percent of the population. Knowing ahead of time a person's Rh factor makes it possible to avoid incompatible transfusions and to correct any incompatibility by a blood transfusion either in the womb or directly after birth. Landsteiner's discovery of blood types has made blood transfusions routine and safe, and has saved the lives of millions of people.

[*See also* **Blood; Rh Factor**]

person A but not from person B, while another sample might clump blood from person B but not from person A. Another might clump both, and yet another might clump neither. Instead of giving up in the face of what seemed chaos, Landsteiner kept at his experiments and data-gathering and eventually saw that a real pattern existed in all of this. From his observations he came up with the idea of mutually incompatible blood groups, which he finally was able to sort out into four groups he called A, B, O, and AB. Landsteiner explained that blood contained certain antibodies that triggered a clumping reaction when one group, or type of blood, was mixed with another. He then showed that blood transfusions were possible if blood was "typed" and if the right type of blood was given to the right patient. Guided by Landsteiner's work, the first successful blood transfusions were achieved at Mt. Sinai Hospital in New York in 1907. Thereafter, Landsteiner's achievement saved many lives on the battlefields of World War I (1914–18), where transfusions of "compatible" blood were first performed on a large scale.

Although Landsteiner continued to work on antibodies, he turned his attention to studying the disease called polio (a viral disease that attacks the nervous system) and was able to show that it was not caused by a bacteria, but was instead traceable to a virus. In the 1920s, Landsteiner joined the Rockefeller Institute for Medical Research in New York and became an American citizen. Although officially retired by 1939, he kept working and, in 1940, discovered yet another blood factor that came to be called the Rh factor (named after the Rhesus monkeys in which it was first discovered). The Rh factor was shown to be responsible for a disease that occurred when mother and fetus have incompatible blood types and the fetus is injured or killed by the mother's antibodies. Landsteiner's brilliant work on blood groups has had a major impact on medicine and health, making life-saving blood transfusions possible. His work on blood typing also is regularly applied in legal and criminal cases in which blood is used as evidence. Landsteiner never really stopped working and died after suffering a heart attack at his laboratory bench.

Botany

Botany is the scientific study of plants. While early humans were very knowledgeable about identifying harmful and beneficial plants, the ancient Greeks were the first to study plants scientifically (or for the sake of gaining knowledge rather than any practical purpose). The study of plants greatly expanded with the seventeenth-century invention of the microscope and today, modern botany uses a whole range of tools that investigate plants at their genetic level.

Plants are multicelled organisms that live by making their own food using the process of photosynthesis to harness the energy of sunlight and convert it into food. Plants are essential to all living things since they provide food, oxygen, energy, and even wood. Since early humans were hunter-gatherers before they learned how to farm, it was especially important to know which plants were good to eat and which were not. Some plants were sometimes found to have medical uses. Until the Greeks studied plants in the fourth century, knowledge of plants consisted primarily of the following types of practical information—which plants were safe, which were harmful, and which were good to cure illness.

THEOPHRASTUS CONSIDERED THE FOUNDER OF BOTANY

The Greek scholar Theophrastus (c.372–c.287 B.C.) began to study plants as life-forms worthy of study by themselves rather than sources of food or drugs. His work, titled *Enquiry into Plants,* survives today and has earned him the title "founder of botany." In this work he studied a wide range of plants and discussed seeds, budding, and the effects of dis-

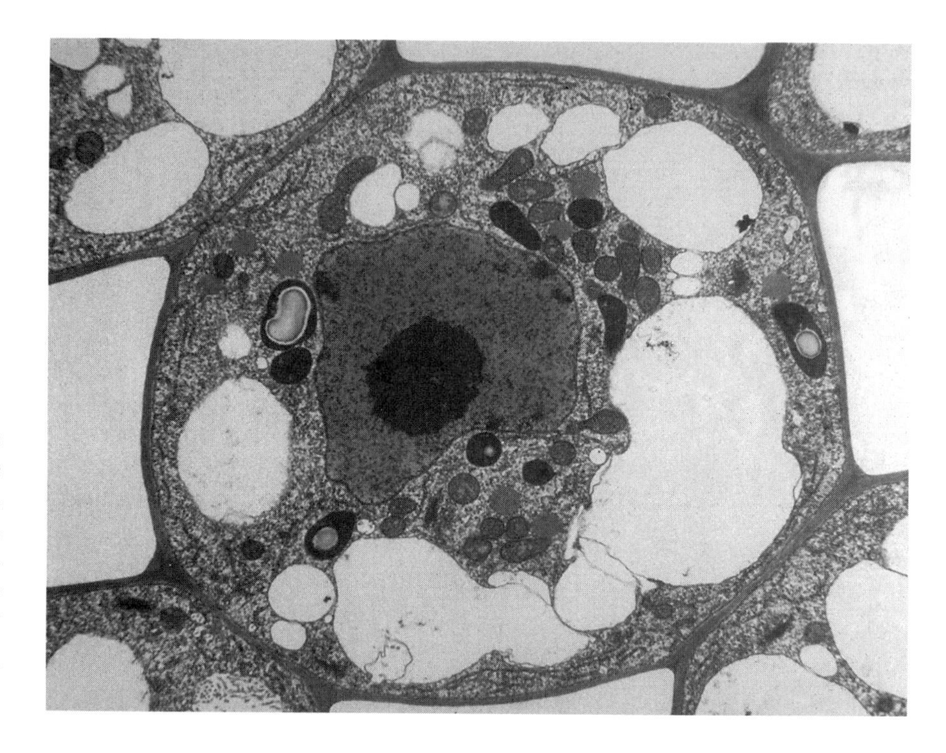

The magnification of a plant cell showing the nucleus, chloroplasts, mitochondria, cytoplasm, vacuoles, and cell wall. (©Photographer, The National Audubon Society Collection/ Photo Researchers.)

ease and weather. He also attempted a classification of plants and described their different parts. Following Theophrastus, other Greeks and Romans were more interested in the practical aspects of plants and most books written were "herbals" or works that contained mainly medicinal (and sometimes mythical) information on plants. The tradition of herbals, which stressed plants that were useful to people, continued throughout medieval times and the Renaissance.

OTHERS EXPAND ON THEOPHRASTUS'S WORK

By the middle of the sixteenth century, the scientific tradition begun by Theophrastus was revived in Germany as several naturalists began to produce botanical books that were based on facts rather than on the elaborate and sometimes fantastic claims of the herbals. These naturalists also began to investigate plant anatomy (the structure or parts of plants) and plant physiology (the internal life processes that take place). This scientific tradition was reinforced in the next century by the invention of the microscope, which allowed a better view of a plant's minute parts. By the end of the seventeenth century, plant anatomy was being studied seriously and many correct scientific discoveries were being made. In 1682, the English physician Nehemiah Grew (1641–1712) published *The Anatomy of Plants* in which he displayed eighty-three full-page plates of microscopic sections of plant stems and roots. Grew was the first to state that flowers contained a plant's sexual reproductive organs. His work and that of others led to the landmark work of the Swedish clergyman and naturalist, Carolus Linnaeus (1707–1778). After traveling through much of Europe studying plants, Linnaeus published in 1735 his *System of Nature* in which he created the modern form of systematic classification known as the binomial (two name) system. The first name he used was the genus, or the type of group, to which they belonged. This was followed by the species, or the particular name. His system soon became useful in classifying all living things. Following Linnaeus, botanists became increasingly specialized and nineteenth-century botany became best known for its discovery of photosynthesis and of the cellular structure of plants. By 1900, the earlier work of the Austrian monk Gregor Mendel (1822–1884), in which he worked out the actual laws of inheritance based on his work breeding pea plants, became well known and modern botany truly began.

Today, the study of botany has many interconnected branches. The major areas of investigation are plant anatomy or the study of the internal arrangement of plant parts; plant physiology or the processes (like photosynthesis) that take place inside a plant; plant morphology or the

THEOPHRASTUS

Greek botanist Theophrastus (327 B.C.–287 B.C.) is considered the father of botany, the scientific study of plants. He was the first to study plants solely for their own sake and not just to learn how they might be put to some practical use. Although few of his writings remain, what did survive became the principle source of botanical information for centuries.

Theophrastus was born on the Greek island of Lesbos and was lucky to study as a very young man with the great Greek philosopher, Plato (c.427 B.C.–c.347 B.C.). After Plato's death, Theophrastus met and became a lifelong friend of the second great philosopher of the ancient world, Aristotle (384 B.C.–322 B.C.). In fact, it was Aristotle who gave him his nickname "Theophrastus," meaning "divine speech." Aristotle had founded a school called the Lyceum which Theophrastus took over after Aristotle's retirement. Under Theophrastus's leadership, the school reached its highest point. There, he carried on Artistotle's teachings in biology, although he concentrated on the study of plants (botany), where Aristotle had specialized in the study of animals (zoology).

Although Theophrastus is believed to have written a great deal on many different subjects, only a small portion of his botanical work survived. In these works, he covered every major aspect of plants—their description, classification, and distribution as well as how plants propagate (reproduce). Significantly, he described the formation of the plant in the seed as being like the fetus of an animal, something produced by it but not a part of it. He identified and grouped more than five hundred species and varieties of plants from those he knew, as well as those from neighboring lands. He classified plants into trees, shrubs, undershrubs, and herbs and

external or visible arrangement of a plant's parts; plant taxonomy or the identification and classification of plants; plant cytology or the study of plant cells; plant genetics or the study of plant inheritance; plant ecology or the study of a plant's relationship to its surroundings; plant paleobotany or the study of fossil plants; dendochronology or tree-ring dating; and ethnobotany or the relations between humans and plants (especially the identification of plants with medical properties).

[*See also* **Plant Anatomy; Plant Hormones; Plant Pathology; Plant Reproduction; Plants**]

developed a way of naming plants based on their external and internal parts, which he called organs and tissues. He also described sexual reproduction in flowering plants, as well as seed germination (when a seed starts to grow and puts out a root) and development. Although the real function of pollen (dustlike grains that contain the plant's male sex cells) was not understood, he wrote detailed descriptions of how to pollinate certain fruit-bearing trees. His knowledge of plants was such that he knew that some flowers bear petals while others do not, and that there were major differences in the seed structure of flowering plants (called angiosperms) and cone-bearing trees (called gymnosperms). In fact, he is credited with inventing the term "gymnosperm" which in Greek means "naked seed." Finally, he described how Greek farmers used certain bean crops to enrich the soil. Today, farmers know that important nitrates (salts from nitric acid) are formed by bacteria that live on the roots of these bean plants, and that they add important nitrogen (a nonmetalic element) to the soil.

Theophrastus is rightly considered to be the founder of botany. Unlike many who followed him, his plant study was focused on learning about plants not for their practical uses (which are many and important), but from a purely scientific aspect, simply in order to learn more about them. His one surviving botanical work contains all the essentials of what today is considered scientific botany. He observed, collected, and systematized his botanical information, and wrote in a clear and accurate manner. Although missing from his work are all of the fabulous folk tales that surrounded plant lore, he brought a scientific mentality to the study of plants. In many ways, Theophrastus was more modern than anyone who followed him for the next two thousand years.

Brain

The brain is the control center of an organism's nervous system. Composed of specialized cells called neurons, the brain receives information from the body's sensory systems. It processes and analyzes the information, and responds by sending out messages that control the rest of the body. All vertebrate (an animal with a backbone) brains are organized and divided into three parts.

Not all organisms have a brain. Simpler forms of life, such as invertebrates (animals without a backbone) like the sponge, jellyfish, and sea anemone have what are called "nerve nets." Animals like earthworms and grasshoppers have a larger collection of cells called "cerebral ganglia."

These collections of nerve cells function as a primitive brain and perform the basic operations of receiving sensory information and acting upon it. More complex forms of life, and certainly all vertebrates, have a distinct and separate organ known as the brain. Described as the control center of the body, the brain's functions are usually described by words like organizing, coordinating, supervising, and governing.

HOW THE BRAIN WORKS

The above words are appropriate in referring to what the brain does, since unlike any other organ, the brain performs what might be called higher functions. For example, no other organ in the body is responsible for the entire organism. It is the brain that receives input, or information, from numerous sensory systems and analyzes and combines that information in order to issue commands to other body systems. The brain regulates and controls other organs and makes sure that the body's "automatic" operations are running automatically. It is the brain that controls

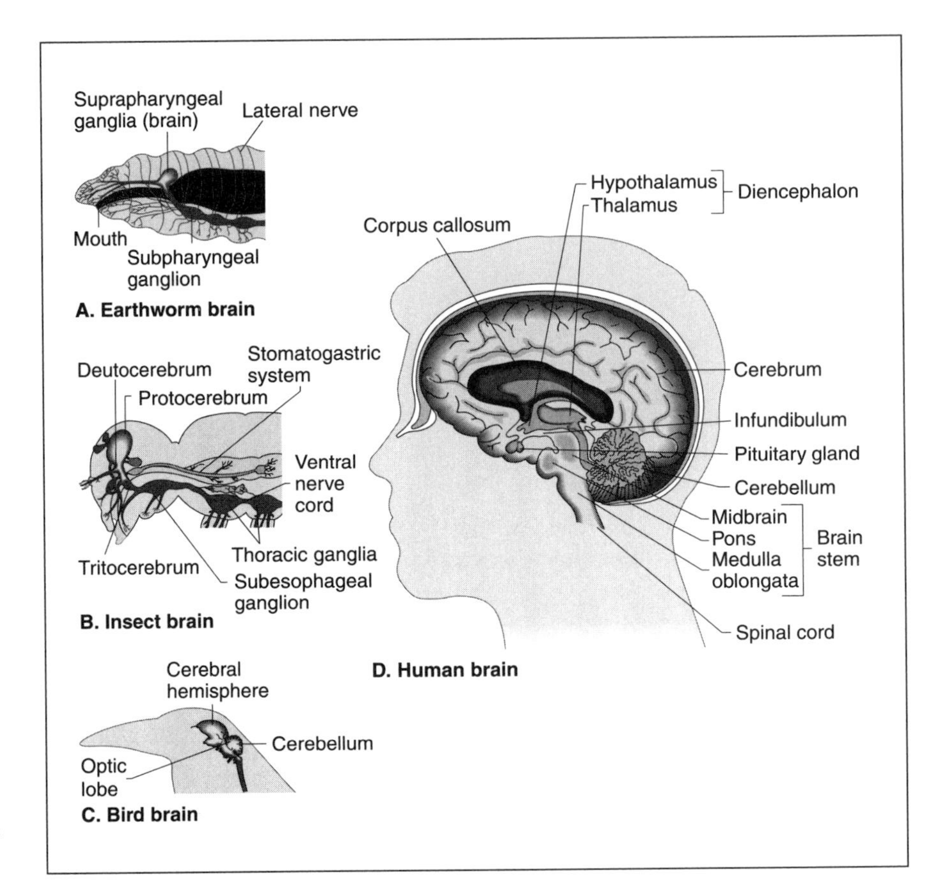

An illustrated comparison of the brains of an earthworm, an insect, a bird, and a human. (Illustration by Hans & Cassidy. Courtesy of Gale Research.)

all of the body's movements. In humans, it is the brain that is responsible for our thought, language, and awareness or consciousness.

Each brain is able to accomplish these complex tasks because it is composed of neurons, or nerve cells. Neurons are the building blocks of the animal brain. Like any cell, neurons have a nucleus (a cell's control center), cytoplasm (a jelly-like substance in a cell), and a membrane. Yet they also have structures not found in any other cells. These structures are long, thin fibers, or threads, called axons and dendrites that extend out from the cell's body. It is these fibers that allow a neuron to receive and send electrical impulses or signals. Axons do the sending and dendrites do the receiving.

PARTS OF THE BRAIN

Although the brain is astoundingly complex, the vertebrate brain can be described in simple terms. All brains can be divided into three main parts: the hindbrain, the midbrain, and the forebrain. The hindbrain, or brain stem, is the most primitive part of the brain and consists of the medulla, the pons, and the cerebellum. Located in the rear of the brain, these regulate several of the autonomic bodily functions, like heartbeat and breathing, and coordinate our movements. The midbrain does a great deal of relaying of information and is responsible for much of the sensory data obtained by the eyes and ears.

The forebrain, in the front, contains what are referred to as the higher brain centers. The cerebrum, the site for our thinking, reasoning, and language, is located in the forebrain. In humans, the cerebrum is divided into left and right hemispheres or sides. Each side is divided again into four lobes. The occipital lobe receives and analyzes visual information. The temporal lobe deals with memory, hearing, and some language functions. The frontal lobe regulates movement and handles language production. The parietal lobe deals with sensations. These two halves communicate by means of a bundle of axons called the corpus callosum, and each side of the brain controls the opposite side of the body.

In humans, the entire brain is protected by the very hard and thick bones of the skull. Between the brain and the skull are three layers of protective tissue called meninges, and a clear, colorless fluid called cerebrospinal fluid that acts as a shock absorber. The brain also contains a group of connected structures that make up what is called the limbic system. Sometimes called the "emotional brain," this system is made up of the olfactory bulb and the amygdala, which senses smells; the hypothalamus, which controls basic drives like hunger, thirst, and sex; the pituitary gland, that secretes growth and other hormones; the hippocampus, which works for memory; and the thalamus, which coordinates sensory information. Overall, the limbic system is responsible for the basic drives, emotions, and involuntary behavior that are critical for an animal's survival.

The human brain weighs about 3 pounds (48 ounces) and contains about a trillion neurons. It is one of the largest organs in the body and its cells use about 20 percent of the body's oxygen. The human cerebrum (the large rounded structure of the brain occupying most of the cranial cavity) is bigger than all of the other parts of the brain put together. The brain does most of its growing after birth, which is part of the reason that human childhood is so long. This growth continues until about the age of 18, but learning takes place until death, and the brain is constantly "remodeling" itself. In other words, the brain creates and preserves new connections in response to new experiences.

THE BRAIN AND THE FUNCTION OF MEMORY

The phenomenon of memory is still not well understood, but scientists know that the brain has short-term and long-term memory. In the short-term memory, the brain holds on to certain information for only a certain amount of time. In the long-term memory, it stores selected information for many years. The brain also regularly enters an altered state of consciousness called sleep. While there is still much to learn about sleep, it is known that sleep allows the body to rest and repair the daily wear and tear it receives. It somehow also lets the brain "recharge" its batteries.

Brain disorders can be serious, since the brain is such an important organ. Organic diseases like Parkinson's and Alzheimer's disease are a leading cause of death. Mental disorders like dementia and schizophrenia often render individuals incapable of having a normal life. Strokes, in which the flow of blood to the brain is temporarily cut off usually by a clot, disable 500,000 Americans each year. According to the part of the brain impacted by the blockage, the individual can suffer some localized paralysis or disability. If any part of the brain is without blood for even a few minutes, its cells die.

[*See also* **Circulatory System; Nervous System**]

Bryophytes

Bryophytes are nonflowering plants that make up one of the major divisions of the plant kingdom. They are composed of moss, liverwort, and hornwort, all of which are small, simple land plants that live in damp places.

The most widely accepted method of classifying the many types of plants in the kingdom Plantae divides it into ten divisions, one of which is called Bryophyta. Plants in this division are unlike plants in any of the other nine since they make up the nonvascular plants. Unlike typical plants, nonvascular plants lack a transport system made up of tubelike vessels or internal pipelines to move their water and food about. Because of this, these plants live very close to the ground and do not have stems, leaves, or roots.

Bryophytes live in almost any part of the world, and are found in the Arctic as well as the tropics. They may have been among the first land plants, since fossils of bryophytes have been found that date to about 400,000,000 years ago. Since bryophytes do not have a specialized method of moving its food and water from one part of the plant to another, all parts of the plant can absorb water and nutrients directly from the environment. Bryophytes also reproduce differently from most plants by using spores rather than seeds. A spore is similar to a seed in that it has an outer coat that protects its inner reproductive cells. Spores are usually released into the air by fungi (such as mushrooms) and are light enough to travel great distances. If a spore lands in a suitable place, it will germinate (begin to grow or sprout) and produce a new fungus. Bryophytes cannot reproduce without water, which is why most are found in moist places.

Mosses, part of the bryophyte group, are soft and leafy, usually never more than 2 inches (5.08 centimeters) tall. They grow by anchoring themselves by rootlike growths called rhizoids. There are nearly 10,000 species of moss. Liverwort and hornwort also are both bryophytes and have a ribbonlike or flat, leafy shape that grows low to the ground and is anchored like the mosses. Bryophytes do not produce flowers since they use spores to reproduce.

[*See also* **Plants**]

Buds and Budding

A bud is a swelling or undeveloped shoot on a plant stem that is protected by scales. Within the scales is a cluster of overlapping, immature or undeveloped leaves or flower petals that will eventually open and develop into new leaves, flowers, or stems. Budding describes the developmental process by which immature plant tissue inside expands and grows into mature structures like leaves or flowers. Botanists (people who specialize in the study of plants) also call "budding" the type of asexual reproduction that occurs when a plant makes a genetically identical copy of itself (as strawberries do when they send out aboveground runners that form new individual plants).

A plant bud is a complex structure that is basically a tiny shoot packed into a small space. It could also be described as a plant's growing point, for it is from a plant's buds that new growth happens. After a bud is formed, if the growing season is ending, the bud will remain dormant or inactive until conditions are right. With proper conditions (usually the warmer and longer days of spring), the protective scales fold back, the bud bursts, and the shoot begins to grow. If a bud is found at the tip of a stem, it is called a terminal bud. Buds that form along the sides of a stem are called lateral buds.

Dissecting a bud reveals its complexity. Starting at the outside, it is protected by bud scales that protect the delicate tissues inside and help it conserve water. The dormant buds of many woody plants are protected by several tough, overlapping scales. The smallest and least-developed part of the bud is found at its center, and is surrounded by slightly older, larger, and more developed parts that overlap and curve around one another. Within the bud, each bit of tissue is already programmed for differentiation after it starts to grow. This means that each part knows that it will eventually become a certain, specific type of plant tissue—perhaps a flower, a leaf, or a stem. Buds do not grow constantly, having what is called episodic growth, meaning that they grow only when conditions are favorable. In climates where the seasons vary greatly, either because of cold winters or extended droughts, buds remain dormant for long periods of time. They "break" or are stimulated to grow when the temperature warms in the springtime or after rains occur. These condition changes are recognized by the plant, which releases certain hormones known as auxins. The release of certain amounts of these chemicals breaks the dormancy and the buds swell and burst open and budding occurs. Buds can also refer to the individual flower before it blooms.

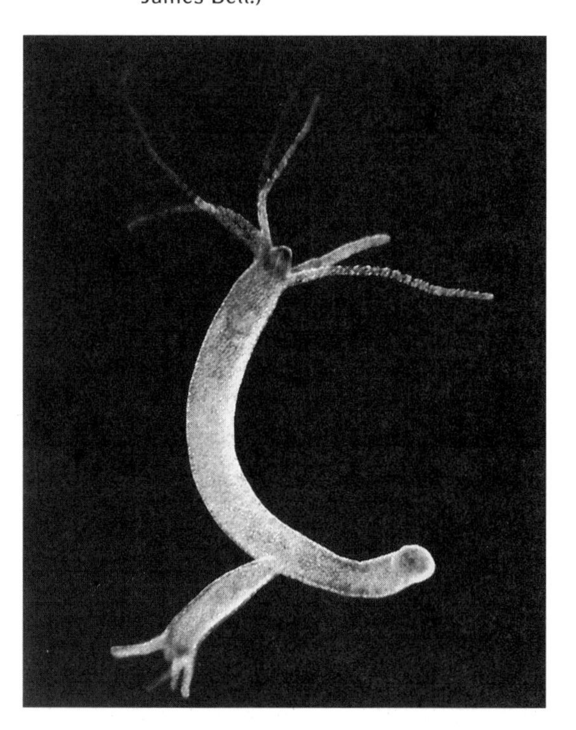

A freshwater hydra in the process of budding. (Reproduced by permission of the National Audubon Society Collection/ Photo Researchers, Inc. Photograph by James Bell.)

Besides describing the developmental process of expansion, growth, and differentiation, budding also can refer to the form of asexual reproduction known as vegetative reproduction. In such cases, certain plants have the ability to duplicate themselves by developing aboveground runners, like strawberry plants do, or belowground roots that send up plantlets, like how grass in a lawn spreads. Other plants like poplar trees put out rhizomes or underground stems that grow outward from the parent. In every case, the new individual plant is genetically identical to the parent. This is unlike sexual reproduction, which produces a genetically unique individual by the mixing of genes. Yeast (which are single-celled fungi) also reproduce by budding as they produce new cells from the parent cells. Some animals reproduce asexually by budding including the many species of hydra, corals, and sea anemones.

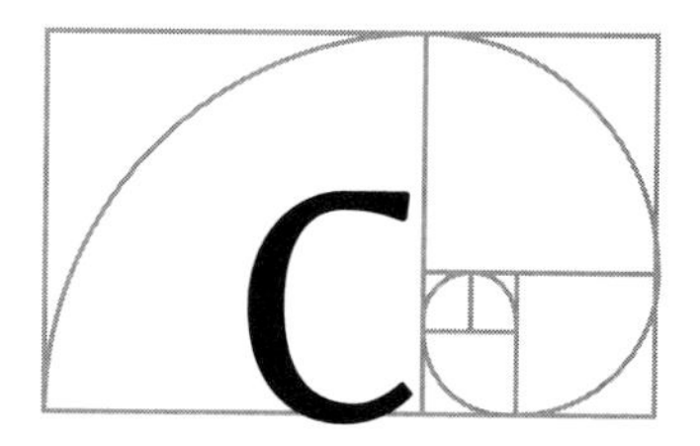

Calorie

A calorie is a unit of measurement of the heat given off by a substance when it is completely consumed. Besides this description, it is also a measurement of the amount of energy in food. The number of Calories in a serving of food tells how much heat energy is available in the food for the body to use.

The word calorie always relates to the notion of energy, although it has two separate uses. The first and older of the two came about when scientists realized the need for accurate measurement, especially the measurement of energy. Once it was discovered that matter is simply energy in another form, scientists realized they would be able to measure how much energy a piece of matter contained if they consumed that matter completely (usually by burning) and somehow measured the energy that was freed. Thus calorimetry, or the measurement of the heat given off by a substance when it is completely consumed in a chemical reaction, was invented. In scientific terms, it was decided that a calorie would be equal to the amount of heat needed to raise the temperature of 0.035 ounces (1 gram) of water by 1 degree Centigrade at standard atmospheric conditions (since things sometimes act differently at high altitudes). This is a very accurate method of measuring potential energy, and it has proven to be extremely useful to many scientific disciplines.

Calorimetry should be distinguished from the other notion of calorie which is used as a measure of the amount of energy in food. This notion of calorie has become so popular and actually so important to good health, that today the word calorie has been replaced by the word "kilocalorie" (also sometimes called a kilojoule). A kilocalorie is exactly what is sounds like: 1,000 calories, since the prefix "kilo" means one thousand times. One kilocalorie is defined as the amount of heat energy needed to raise the temperature of 0.028 ounces (1 kilogram) of water by 1 degree Centigrade. Finally, when calories are written as "Calories" with a capital "C," it means a kilocalorie. In the United States, when someone refers to Calories in foods, we are always talking about kilocalories since the original calorie (1/1,000th of a Calorie) is too small a unit of measurement to be practical when labeling foods.

It has long been known that people need to take in a certain amount of Calories everyday to maintain good health. This number changes quite a bit according to the age and level of activity of an individual. For example, an expectant mother would need to take in more Calories since her system is supporting a developing fetus. In the same way, a person who spends a great deal of time outdoors in a frigid environment needs many more Calories than a person sitting warmly at home, since the former's body must burn more Calories just to maintain steady body heat. The average recommended daily requirement for men is considered to be 2500 Cal and 2000 Cal for women, but these totals should decrease as a person gets older.

Although it is much easier to say than to carry out, if a person wants to maintain his or her weight, he or she should balance intake of Calories with output of energy. Many adults are able to do this with little or no effort, and therefore maintain a consistent body weight over very long periods. Losing weight requires either a reduction in caloric intake or increase in energy output (or both). Many of today's packaged food products provide the buyer with nutrition information, including the amount of Calories in an average serving. It is therefore possible to obtain a fairly good estimate of the amount of Calories consumed during a twenty-four-hour period.

[*See also* **Nutrition**]

Carbohydrates

Carbohydrates are a group of naturally occurring compounds that are essential sources of energy for all living things. Carbohydrates are manufactured by green plants, which use them for energy and to build their cell walls. Animals are able to both use and store the energy that carbohydrates contain. Carbohydrates include such organic compounds as sugars, starches, and cellulose.

Chemically, carbohydrates are natural compounds of carbon, hydrogen, and oxygen and were described as "carbon hydrates" by the chemists who first encountered them (since they mistakenly thought that carbohydrates were simply a compound of carbon and water). We now know that a carbohydrate molecule is made up of equal numbers of carbon and oxygen atoms (the smallest units of elements), with twice as many hydrogen atoms. Almost all the energy consumed by living things comes from carbohydrates that are manufactured by plants.

CARBOHYDRATES PRODUCED BY PHOTOSYNTHESIS

Plants manufacture carbohydrates using the process known as photosynthesis. Because plant cells contain chloroplasts inside their cells, plants are able to use the sunlight absorbed by their leaves that they combine with the water from the soil and the carbon dioxide from the air to make a simple form of sugar called glucose, which is a carbohydrate. Glucose is packed with energy and plants use it to make cellulose (their building

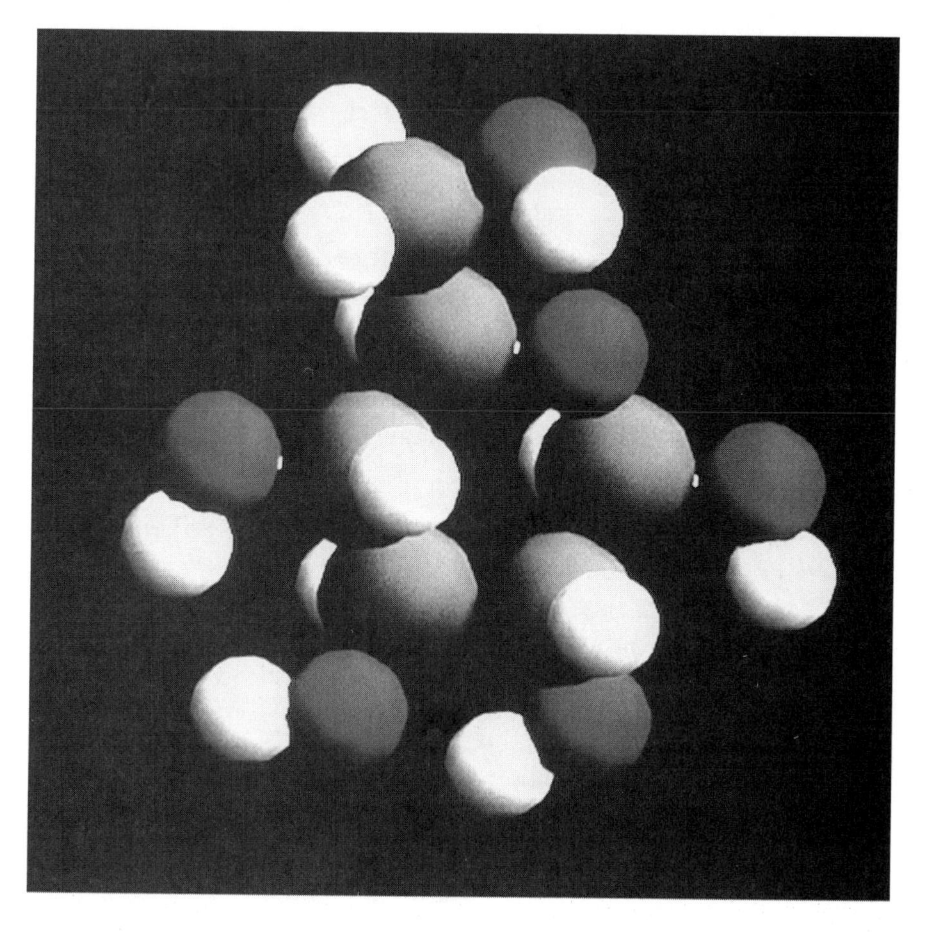

A computer graphic representation of fructose, a common carbohydrate found in many fruits and vegetables. (©Chemical Design Ltd./Science Photo Library, National Audubon Society Collection/Photo Researchers, Inc. Reproduced by permission.)

material). What they do not use, they store as starch in the form of seeds, roots, or fruits. Animals that eat plants take in their carbohydrates and reverse a plant's metabolic process. This occurs by the process of digestion in which animals break down complex carbohydrates into the original glucose that animals are then able to absorb into their bloodstream. The animals' bloodstream then carries the glucose to every cell in the body and cells turn the glucose into energy by a process known as oxidation, in which glucose combines with oxygen and releases energy. In a sort of reverse process, an animal turns a plant into the energy that the plant "captured" from the sun.

TYPES OF CARBOHYDRATES

There are three major types of carbohydrates: monosaccharides, disaccharides, and polysaccharides. As the names imply, each is more complex than the other. Monosaccharides are the simplest of the carbohydrates since they consist of a single carbohydrate unit that cannot be broken down into anything simpler. The three most common sugars in this group are glucose, fructose, and galactose. Glucose is found in fruits and vegetables; fructose is found in honey and some fruits and vegetables; and galactose is found in milk. Disaccharides are more complex since they consist of two joined monosaccharides. The three most nutritionally important disaccharides are sucrose (ordinary table sugar), maltose (found in sprouts), and lactose (found in the milk of mammals). Polysaccharides are made up of ten or more monosaccharides and are often highly complex chains of long molecules. Carbohydrates are stored in the form of polysaccharides (as starch in plants and as glycogen in animals).

THE IMPORTANCE OF CARBOHYDRATES

Much has been learned about the role carbohydrates play in our diet and health. For instance, it was once believed that avoiding carbohydrates like bread, potatoes, and pasta would keep a person slim. Scientists now know that this is not the case since not all carbohydrates are created equal and that the type of carbohydrate eaten is more important than the total amount consumed. Low-fiber carbohydrates are digested very quickly and behave like sugar, giving a quick peak in energy. On the other hand, high-fiber carbohydrates like whole grains are digested more slowly, providing a steady and gradual source of energy. Athletes have learned that continuous exercise, such as running a marathon, can burn up all of the body's supply of glycogen in about two or three hours. After that the body must start converting its stored fat into glucose. Since this is a fairly slow process, the athlete operating on stored fat becomes fatigued. To overcome this, some athletes whose competition lasts longer than ninety minutes began the practice of "carbohydrate loading" before competition. Beginning a week before competition, athletes consume an increasing amount of carbohydrate-rich foods daily.

The importance of carbohydrates to all living things cannot be overemphasized since it is the primary source of energy for plants and animals. We have learned that "starchy foods," such as potatoes, which were once thought to be inadequate for humans in large amounts, are in fact a necessary part of every meal. The U. S. Food and Drug Administration now recommends that people's carbohydrate consumption should be increased and fat consumption should be reduced.

[*See also* **Fruits; Photosynthesis**]

Carbon Cycle

The carbon cycle is the process in which carbon atoms are recycled over and over again on Earth. Carbon recycling takes place within Earth's biosphere (the region of the Earth that supports life) and between living things and the nonliving environment. Since a continual supply of carbon is essential for all living organisms, the carbon cycle is the name given to the different processes that move carbon from one organism to another. The complete cycle is made up of "sources" that put carbon back into the environment and "sinks" that absorb and store carbon.

Earth's biosphere can be thought of as a sealed container into which nothing new is ever added except the energy from the sun. Since new matter can never be created, it is essential that living things be able to reuse the existing matter again and again. For the world to work as it does, everything has to be constantly recycled. The carbon cycle is just one of several recycling processes, but it may be the most important process since carbon is known to be a basic building block of life. Carbon is the basis of carbohydrates, proteins, lipids, and nucleic acids—all of which form the basis of life on Earth.

Since all living things contain the element carbon, it is one of the most abundant elements on Earth. The total amount of carbon, whether we are able to measure it accurately or not, always remains the same, although carbon regularly changes its form. A particular carbon atom located in someone's eyelash may have at one time been part of some now-extinct species, like a dinosaur. Since the dinosaur died and decomposed millions of years ago, its carbon atoms have seen many forms before ending up as part of a human being. It may have been part of several plants and trees, been free-floating in the air as carbon dioxide, been locked away in the shell of a sea creature and then buried at the ocean bottom, or may have been part of a volcanic eruption. Carbon is found in great quantities in Earth's crust, its surface waters, the atmosphere, and the mass of green plants. It also is found in many different chemical combinations, including carbon dioxide (CO_2), calcium carbonate ($CaCO_3$), as well as in a huge variety of organic compounds such as hydrocarbons (like coal, petroleum, and natural gas).

CARBON CYCLE PROCESSES

If a diagram were drawn showing the different processes that move carbon from one form to another, its main processes would be photosynthesis, respiration, decomposition, combustion of fossil fuels, and the natural weathering of rocks.

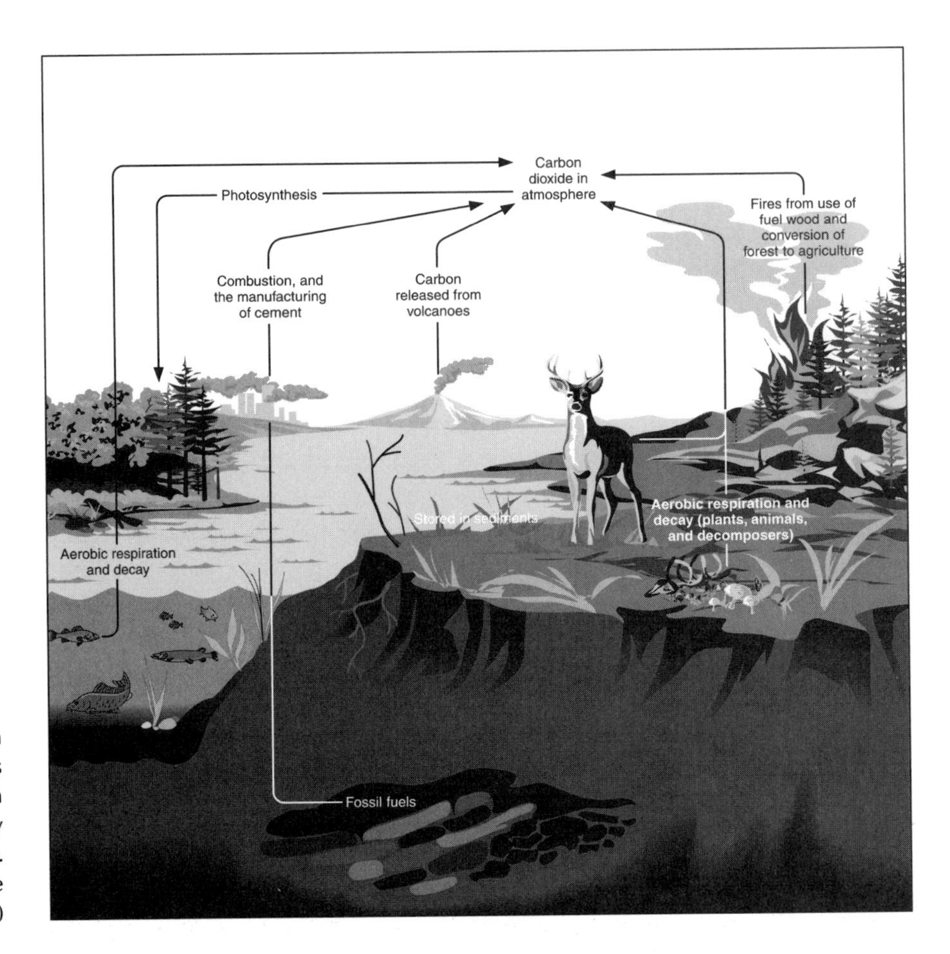

An illustrated diagram showing the processes found in the carbon cycle. (Illustration by Hans & Cassidy. Courtesy of Gale Research.)

PHOTOSYNTHESIS

Carbon exists in the atmosphere as the compound carbon dioxide. It first enters the ecological food web (the connected network of producers and consumers) when photosynthetic organisms, such as plants and certain algae, absorb carbon dioxide through tiny pores in their leaves. The plants then "fix" or capture the carbon dioxide and are able to convert it into simple sugars like glucose through the biochemical process known as photosynthesis. Plants store and use this sugar to grow and reproduce. When plants are eaten by animals, their carbon is passed on to those animals. Since animals cannot make their own food, they must get their carbon by eating plants or by eating animals that have eaten plants.

RESPIRATION

Respiration is the next step in the cycle, and unlike photosynthesis, it occurs in plants, animals, and decomposers. Although we usually think of breathing oxygen when we hear the word "respiration," it has a broader meaning that involves oxygen. To a biologist, respiration is the process in which oxygen is used to break down organic compounds into carbon dioxide (CO_2) and water (H_2O). For an animal, respiration includes taking in oxygen (and releasing carbon dioxide) and oxidizing its food (or burning it with oxygen) in order to release the energy the food contains. In both cases, carbon is returned to the atmosphere as carbon dioxide. Carbon atoms that started out as components of carbon dioxide molecules have passed through the body of living organisms and been returned to the atmosphere, ready to be recycled again.

DECOMPOSITION

Decomposition is the largest source through which carbon is returned to the atmosphere as carbon dioxide. Decomposers are microorganisms that live mostly in the soil but also in water, and which feed on the rotting remains of plants and animals. It is their job to consume both waste products and dead matter, during which they return carbon dioxide to the atmosphere by respiration. Decomposers not only play a key role in the carbon cycle, but break down, remove, and recycle what might be called nature's garbage.

HUMANS INCREASE CARBON DIOXIDE IN THE ATMOSPHERE

In recent history, humans have added to the carbon cycle by burning fossil fuels. Ever since the rapid growth of the Industrial Revolution when people first harnessed steam to power their engines, human beings have been burning carbon-containing fuels like coal and oil for artificial power. This constant burning produces massive amounts of carbon dioxide, which are released into Earth's atmosphere. Fossil fuel consumption could be an example of a human activity that affects and possibly alters the natural processes (photosynthesis, respiration, decomposition) that nature had previously kept in balance. Many scientists believe that carbon dioxide is a "greenhouse gas." This means that it traps heat and prevents it from escaping from Earth. As a result, this trapped gas leads to a global temperature rise, which can have disastrous effects on Earth's environment.

Not all carbon atoms are always moving somewhere in the carbon cycle. Often, many become trapped in limerock, a type of stone formed on the ocean floor by the shells of marine plankton. Sometimes after millions of years, the waters recede and the limerock is eventually exposed to the elements. When limerock is exposed to the natural process of weathering, it slowly releases the carbon atoms it contains, and once again they become an active part of the carbon cycle.

[*See also* **Biosphere; Carbon Dioxide; Decomposition; Photosynthesis; Respiration**]

Carbon Dioxide

Carbon dioxide (CO_2) is, along with oxygen and nitrogen, one of the major atmospheric gases. Although it only makes up .03 percent of the atmosphere, it is vitally important for all living things. A colorless and odorless gas, carbon dioxide is made up of a central carbon atom joined to two oxygen atoms.

Early scientists were able to observe the effects of carbon dioxide long before they knew exactly what it was. At the start of the seventeenth century, the common air humans and animals breathe was thought to be a single substance or element. However, around 1630, a Flemish physician, Jan Baptista van Helmont (1577–1644), became the first to discover that there were other "vapors" that were different from ordinary air. He coined the word "gas" to describe the vapors that were given off when wood was burned (actually carbon dioxide). He also recognized that this same gas was produced by the process of wine fermentation. In 1756, the Scottish chemist, Joseph Black (1728–1799), proved that carbon dioxide is present in the atmosphere and that it combines with other chemicals to form compounds. Black also learned that the gas he called "fixed air" was present in exhaled breath and that it was heavier than ordinary air. It was also known by then that a candle flame would eventually go out when enclosed in a jar with a limited supply of air and that a small bird would die in the same tight jar. Near the end of the eighteenth century, chemists finally began to realize that gases were important and should be weighed and accounted for whenever chemists analyzed chemical compounds, as was done with solids and liquids. This more scientific method and approach would eventually lead to a real understanding of the nature and role of carbon dioxide.

Following the 1771 discovery by English chemist Joseph Priestley (1733–1804) that plants give off oxygen (which he called "dephlogisticated air"), the Dutch physician Jan Ingenhousz (1730–1799) demonstrated in 1779 that during photosynthesis (the plant process that uses sunlight to make food), green plants take in carbon dioxide (as well as give off oxygen). He therefore established the key fact that plants, in the presence of light, consume the carbon dioxide produced by humans and animals and give off the oxygen which is in turn consumed by both humans and animals.

In the bodies of humans and animals that breathe air, carbon dioxide is generated by the cells as a waste product (similar to the release of carbon dioxide when wood burns). The lungs remove it from the body, which received it from the bloodstream. The amount of carbon dioxide in the blood affects how humans and animals breathe; a rise stimulates the rate of breathing and a drop lowers it. When air consists of 3 percent of carbon dioxide, it is difficult to breathe. If the amount goes above 10 percent, a loss of consciousness occurs. At about 18 percent humans and animals suffocate. Although this rarely happens, in 1986 a huge cloud of carbon dioxide suddenly exploded from Lake Nyos in northwestern Cameroon (Africa) and suffocated more than 1,700 people and 8,000 animals. It was later learned that atmospheric conditions probably caused the colder water at the bottom of the lake to turn over and quickly release a large amount of carbon dioxide at the surface.

Increasingly, we are becoming aware of the critical role carbon dioxide plays in the health of the global environment. The amount of carbon dioxide in the atmosphere affects our climate since it acts to trap the heat escaping from Earth's surface and reflect it back down. Many scientists argue that when there is too much carbon dioxide in the atmosphere, a "greenhouse effect" will occur and cause the planet to overheat. There is no doubt that atmospheric levels of carbon dioxide have increased over the years as a result of massive amounts of forests that would consume carbon dioxide being destroyed. Also, the amount of carbon dioxide in the atmosphere has also increased because of the steady burning of fossil fuels (which releases carbon dioxide as a by-product). Fermentation and the breakdown of organic matter also play a role toward increasing the amount of carbon dioxide in the atmosphere. Many argue that Earth's temperature could be raised by these influences to a level at which all life is threatened.

Despite these dire predictions, recent scientific calculations of the amount of carbon dioxide in the atmosphere suggest that there is significantly less than should be expected. So far, scientists have not been able to account for the "missing" carbon dioxide. Many think that we may be underestimating the amount or rate (or both) of photosynthesis that takes place on a global scale. If that is the case, Earth is in better shape than first thought.

[*See also* **Carbon Cycle; Carbon Family; Greenhouse Effect; Photosynthesis; Pollution**]

Carbon Family

Carbon is a chemical element that is essential to all life on Earth. Since it can link up with the atoms of other elements to form an immense variety of stable compounds, carbon is able to form the chemical structures basic to all life. More than 1,000,000 compounds have been identified in the family of carbon compounds.

Carbon has been called the "backbone" element of life. This is because carbon is such an unusual and versatile element that it forms more compounds than any other element. An element is a pure substance that contains only one kind of atom. There are more than ninety different chemical elements on Earth, and less than one third of them seem to be essential to life. Of these, only oxygen, hydrogen, nitrogen, and carbon are needed in any real quantity to sustain life. Of these four, only carbon has the ability to combine so easily with other elements that it forms a part of every compound important to life. A single carbon atom can make four bonds, or link up with four other atoms. These can be other carbon atoms, atoms of other elements, or a combination of the two. The huge variety of compounds it forms is explained by the fact that one carbon atom can easily join with another and form long, stable chains. These can be extended indefinitely, so that if only a few other elements are added, the number of compounds it is possible to make becomes almost limitless.

In its pure form, carbon exists only as diamonds and graphite. Both are a form of crystal, although in the diamond (the hardest known mineral), each carbon atom is linked to four others, therefore forming a solid structure. However, in graphite (a soft, almost greasy black solid) the carbon atoms are linked in such a way that they form layers that slide over each other like plates. Aside from diamonds and graphite, most carbon occurs in combination with other elements, and in nearly all of its forms it is linked with, or was once part of, a living organism. In fact, of the nine characteristics of all living things, the presence of carbon is usually listed as the first. In human bodies, carbon, oxygen, and hydrogen are the three most abundant elements, as they represent 93 percent of its weight. Structurally however, carbon is the most important element in the body.

The carbon family is really a family of compounds. All animals and plants consist of carbon compounds, some of which are groups of small organic molecules, while others are large organic molecules called macromolecules. By definition, an organic compound is a compound that contains carbon. In living things, there are four main types of organic compounds—carbohydrates, lipids, proteins, and nucleic acids. An example of small molecules is table sugar, which is a carbohydrate and therefore an example of a carbon compound. Nucleic acids are examples of macromolecules and are the most complex carbon-containing substances in living things.

As a basic building block of life and one of the most abundant elements on Earth, carbon is recycled again and again. This is necessary

FRIEDRICH KEKULE

German chemist Friedrich Kekule (1829–1896) was the first to discover the nature of the carbon atom (a unit of matter) and to explain how carbon atoms arrange themselves. This allowed him to understand the structure of organic compounds and to lay the groundwork for the modern structural theory in organic chemistry.

Friedrich Kekule was born in Darmstadt, Germany, and entered the University of Giessen intending to become an architect (a person who designs buildings). There, however, he sat in on the lectures of the great German chemist, Justus von Liebig (1803–1873), and switched immediately to chemistry. He eventually obtained his doctorate in chemistry and began to teach and do research. His early interest in architecture may have influenced the chemicals problems he studied, since from the beginning, he was always interested in the actual structure of chemical compounds. Until the mid-1850s, chemists would give the atomic makeup of a compound such as water (H_2O) by listing the numbers of each element in a certain order. Therefore, H_2O meant that water contained two hydrogen atoms and one oxygen atom. They did not, however, state exactly what the arrangement or structure of these atoms were. In other words, which atom was bonded or linked to which, and in what order? Many chemists doubted the value of even knowing this, while others believed that their structure might not even be knowable.

In 1858, Kekule decided that structure was important and that it was knowable. Kekule felt that such fixed combinations of elements, like H_2O, had a basic structure all their own and that they could be represented as a pattern of atoms linked in a certain order. Kekule then focused on carbon, an element found in all living things, and argued that any one carbon atom would always form four bonds, no more and no less, in a compound. He then also suggested that carbon atoms had the ability to bond with each other in endless chains. Therefore, a single carbon atom might bond to four other carbon atoms. When Kekule adopted the suggestion by the Scottish chemist Archibald Scott Couper (1831-1892) that a dash or dotted line be used to represent the chemical bond, Kekule was then able to illustrate the structure of water (H_2O) as H-O-H.

Kekule's structural formulas allowed chemists to make sense out of organic compounds (compounds that contain carbon), since he showed that it was the structural arrangement of the atoms that made a compound what it was. For example, the compound known as ethyl alcohol has six hydrogen atoms, two carbon atoms, and one oxygen atom. However, the very different compound known as dimethyl ether has the exact same number of elements (six hydrogen, two carbon, and one oxygen atom), yet a slightly different arrangement of the oxygen atom gives the compound an entirely different nature. Kekule's structures proved to be the key to understanding organic compounds, and actually laid the foundation for the structural theory of organic chemistry.

Kekule's second great achievement was his discovery of the structure of the benzene (a clear, colorless, flammable liquid used to make insecticides) molecule. Chemists knew that its formula was C_6H_6, but no one could arrange these atoms in the proper way. The story goes that in 1865, Kekule fell into a half-sleep while still thinking about the benzene problem, and he dreamt that he saw atoms dancing wildly. Suddenly he saw the tail end of one chain attach itself to the front end, and awoke knowing that he had solved the benzene problem. He then immediately drew a structure that had six carbon atoms arranged in a circle, joined by alternative single and double bonds, with a single hydrogen atom attached to each carbon atom. This is a favorite story of many science historians, showing how Kekule's dreamlike state, in which he saw something like a snake devouring its tail, influenced him to make a real scientific breakthrough. Kekule's work on structure is valid today, and still helps chemists to both show what an organic compound looks like and to predict its reactions.

since the total amount of carbon on Earth always remains the same. However, its form changes according to what part of the carbon cycle it is in. For example, the same carbon atom may be absorbed from the atmosphere by a plant that takes in carbon dioxide. It may then become part of an animal who eats the plant, then part of the atmosphere when the animal breathes, or part of the soil when the animal dies, and so on.

[*See also* **Carbon Dioxide; Carbon Monoxide**]

Carbon Monoxide

Carbon monoxide is an odorless, tasteless, colorless, and poisonous gas. Although most of the carbon monoxide in the atmosphere comes from natural sources, a great deal is also added by the burning of fossil fuels by automobiles and industry. It is an extremely poisonous gas if inhaled, since it kills by preventing oxygen from reaching the cells.

Carbon monoxide (CO) is a compound consisting of one carbon atom (the smallest unit of an element) and one oxygen atom. It is normally in the air people breathe, although in extremely small amounts. It is produced naturally when a substance that contains carbon decays or breaks down in the absence of oxygen. This happens in swamps where there is little oxygen. Carbon monoxide is also produced in greater quantity when carbon-containing substances are burned without enough oxygen being present. This often happens when gas furnaces, wood stoves, and space heaters malfunction or are not properly vented. Automobile engines and other gasoline or diesel-powered motors also generate carbon monoxide. The exhaust from these motors can be deadly if they are operated in enclosed areas or attached garages.

Carbon monoxide can kill a person who breathes it. It does this by preventing the blood from being able to carry oxygen. Without oxygen, people and animals soon die. Once inhaled, carbon monoxide combines with the hemoglobin (the oxygen-carrying substance in the blood) to the exclusion of oxygen. In fact, carbon monoxide combines with hemoglobin two hundred times faster than oxygen does. Additionally, the hemoglobin does not release the carbon monoxide as it does the oxygen. Thus, as more and more red blood cells pick up carbon monoxide, the total number available to deliver oxygen to the cells keeps decreasing, and soon the person slowly falls into a sleeplike state. Deprived of oxygen, the brain begins to slow its functions. Eventually, bodily functions stop, and the person dies.

This oxygen deprivation does not always kill immediately. Low-level exposure can cause flu-like symptoms including shortness of breath, mild headaches, fatigue, and nausea. Higher levels may cause dizziness, severe headaches, mental confusion, and fainting. Prolonged exposure can cause death. According to the U. S. Consumer Product Safety Commission, more than 2,500 people will die and 100,000 will be seriously injured by carbon monoxide poisoning over the next 10 years.

People who smoke also run the risk of harming themselves with this toxic gas. The carbon monoxide present in cigarette smoke not only excludes oxygen from binding to hemoglobin, but it also prevents it from picking up carbon dioxide (which is the waste product of breathing). This means that the person's heart has to pump harder to try and rid the body of carbon dioxide wastes.

Large cities had an especially bad problem in the past as unhealthy amounts of carbon monoxide would build up during rush hour because of automobile exhaust. Newer cars are now equipped with catalytic converters that chemically change carbon monoxide into harmless carbon dioxide. Also, in the last decade or so, carbon monoxide detectors for the

Victims of carbon monoxide poisoning breathe oxygen at a hospital in Philadelphia. (Reproduced by permission of AP/Wide World Photos.)

home have become practical, and many cities are now requiring that at least one carbon monoxide detector be installed in every home, apartment, and hotel. Since carbon monoxide is odorless, tasteless, and colorless, these monitors are invaluable in alerting people to possible exposure.

[*See also* **Carbon Family**]

Carnivore

The term carnivore is thought by many to refer to any meat-eating organism, but in the life sciences it is applied to a certain family of mammals (Carnivora) that have specially shaped teeth and live by hunting. Carnivores are animals that obtain most of their nutrition from eating other animals, and their name comes from a combination of Latin words that literally mean "flesh devourers." Carnivores are always at the top levels of every food chain.

CHARACTERISTICS OF CARNIVORES

There are eight families of terrestrial (land-dwelling) carnivores and three families of aquatic (water-based) carnivores. A family is a classification term that includes one or more genera (singular, genus), and a genus contains one or more species. What best defines a carnivore are its teeth, although most have powerful jaws and a keen sense of smell as well. Since it lives primarily by hunting, catching, killing, and eating its prey, a carnivore has teeth that are specially shaped for all of those demanding tasks, especially for gripping and cutting. These specialized teeth come in sets that have particular functions. The canine teeth are the longest. It is with these curved weapons that a carnivore both grabs and punctures its prey. The canine teeth are best used for holding prey since they can pierce it deeply. The chisel-shaped incisors are next to the canines and are at the center of the jaw. With these, the carnivore bites into food and slices it up. Carnivores also have a unique set of teeth called carnassials that form the first set of molars. These sharp teeth have a pointed edge, and the top slides along the bottom like scissors. Carnassials are used to slice through tough flesh and gristle (cartilage especially present in meat) and to crack open bones. Since they are near the hinge of the jaw, they can close with great force. After a carnivore has caught, killed, and eaten its prey, its specialized digestive system takes over. Since its main diet of meat is easier to digest than the tough material of plants (plant cells are surrounded by a tough wall of cellulose), a carnivore's intestines are much shorter than those of a herbivore (plant-eating organisms) like a horse or cow. A carnivore also a much simpler stomach.

As a predatory animal or one that lives by killing and eating other animals, a carnivore has other distinguishing characteristics that enable it to capture and subdue its prey. One of these is its brain, which is often fairly large and complex. Such a brain means that a carnivore's behavior can be somewhat flexible and that it can rear its young. Some carnivores are particularly social creatures and hunt in packs. This allows them to overwhelm creatures that an individual could not catch on its own. Most carnivores are either speedy runners or very quick and nimble. For example, the cheetah is the fastest land animal on Earth, able to reach 75 miles-per-hour (120.68 kilometers-per-hour) in short bursts. Others have endurance like wild African dogs that can run for 3 miles (4.83 kilometers) at a speed of 35 miles-per-hour (56.32 kilometers-per-hour). Because carnivores must find and catch their food, they are often very active and aggressive animals. However, since they do not have to eat continuously the way herbivores do, they are able to spend more time relaxing in between meals.

LAND-DWELLING AND SEA-DWELLING CARNIVORES

There are about 240 species of carnivore belonging to two suborders or major groups: Fissipedia, or land-dwelling carnivores, and Pinnipedia, or sea-dwelling carnivores. Some biologists, however, consider Pinnipedia to be a separate order like Carnivora. There are three families that make up Pinnipedia and eight families in Fissipedia, each of which has its own specialties and characteristics.

Mustelidae Family. One of the largest members of the Fissipedia group is the Mustelidae family consisting of small carnivores like skunks, badgers, weasels, and ferrets. These sleek animals can burrow into hard-to-reach places and catch their prey. Some of these carnivores depend on foul-smelling spray for defense. They are aggressive and often take on animals that are larger than they are.

Procyonidae Family. The Procyonidae family includes raccoons and the tropical coati, which are less carnivorous and more omnivorous creatures (eating both plants and animals). Tropical coati have teeth that reflect their diet since their carnassials are more like grinding teeth. They are usually slow runners but excellent climbers and often live in groups.

Canidae Family. The Canidae family is especially diverse, consisting of wolves, foxes, and dogs. Very adaptable animals with a keen sense of smell, they are good at running and often hunt in packs. The dog was probably the first animal to be domesticated (tamed) by humans. They live in groups, take care of their young, and are very territorial (defending where they live).

Felidae Family. The Felidae family is made up of some of the most efficient carnivores: the cats. Divided into two main groups called simply big cats and small cats, they can range in size from a tiny 2-pound (.91-kilograms) animal to one more than 750 pounds (340.50 kilograms). The big cats include lions, tigers, jaguars, leopards, and cheetahs. Among the small cats are bobcats, lynxes, pumas, and domestic cats. An interesting difference between the two groups is that big cats can roar but not purr, while small cats can purr but not roar. Despite this oddity, all cats are especially good hunters, possessing retractable claws that are kept razor-sharp and allow them to pad silently while stalking prey. They have large, pointed canine teeth and forepaws used to swipe at and claw their prey. Finally, most cats are solitary hunters.

Hyaenidae Family. One of the more unusual groups are the members of the Hyaenidae family. Hyenas are skilled hunters who work together in packs and are one of the few carnivores that will regularly eat carrion (an already-dead animal). Hyenas are fiercely wild-looking, with heavy bodies and front legs that are longer than the rear ones, giving them a crouching or lurching-forward look. Their skulls are strong and their powerful teeth can easily crush bone, their favorite food.

Viverridae Family. The Viverridae family is represented by civets and genets which live in tropical areas. They resemble weasels and have long

A pack of spotted hyenas in Tanzania eat a zebra. Hyenas are one of the more unusual groups of carnivores. (Reproduced by permission of the National Audubon Society Collection/ Photo Researchers, Inc. Photography by Robert Caputo.)

noses and short legs. Besides meat, they will also eat insects as well as fruit and eggs, and possess scent glands that produce musk, long-sought for its use in perfumes and lotions.

Herpestidae Family. The Herpestidae family is made of many species of mongoose and is an offshoot from the Viverridae family. Some species are solitary while others are very social and live in groups. Viverridae eat insects as well as meat.

Ursidae Family. Probably the most awe-inspiring and terrifying of the carnivores are the members of the Ursidae family, better known as bears. Including the polar bear, black bear, brown bear, or grizzly bear, all have large bodies and short, powerful limbs. While their sense of smell is much better than their hearing or sight, bears can walk or run upright on the soles of their feet. They are fiercely aggressive when provoked and have few natural enemies. Despite their appearance, bears are mostly omnivorous, and spend more time foraging for insects and berries than catching and eating prey. The exception are polar bears, which are mainly flesh-eaters and are such strong swimmers they have been known to pursue they favorite food (seals) as far as 40 miles (64.36 kilometers) from land.

Phocidae, Otariidae, and Odobenidae Families. The three families of the aquatic Pinnipedia suborder are made up of the Phocidae family (sea lions, fur seals, and earless seals), Otariidae family (eared seals), and the Odobenidae family (walrus). Called pinnipeds, all of these marine animals reproduce on land despite the fact that they spend most of their time in the water. All hunt underwater and have adapted to this environment, having nostrils that can close, limbs that are modified fins, and an insulating layer of fur or blubber to reduce the loss of body heat in cold water. Their hearing is especially keen and they can dive to great depths.

All meat eaters do not belong to the Carnivora order, as there are birds (eagles and hawks), reptiles (snakes and alligators), fish (sharks and barracuda), and even plants (Venus's flytrap) that regularly live on a diet of meat. Although many humans regularly eat meat, they are instead classified as omnivores because of their varied diet.

Cell

The cell is the building block of all living things—basic to their makeup and basic to their functions. All cells also have the same processes: they "breathe" and take in food, get rid of wastes, grow, reproduce, and eventually die. It is a self-contained, living unit that takes in and expends energy. A living organism can be made up of a single cell or trillions of cells.

Life scientists have divided the cells of all living things into two types based on whether they have a distinct nucleus (the control center of the cell) or not. They have named these categories prokaryotes (no nucleus) and eukaryotes (nucleus). Prokaryotic cells are found among the simplest of all living things, like bacteria and algae. Not only do these single-celled organisms lack a distinct nucleus but they are also missing many of the other sophisticated structures that perform specialized functions in the cells of more complicated organisms. Eukaryotic cells describe the cells of nearly every other life forms that have more than one cell. Animal cells and plant cells are therefore both eukaryotic.

As the basic unit of life, cells make up every living thing—from tigers and people to cockroaches and flowers. However, although all living things are made up of cells, their size and shape will vary according to their function. In plants and animals, cells have specialized jobs to do, and in an animal, skin cells are different from muscle cells, which are different from nerve cells. These individual specialized cells group together to form tissues which, in turn, form organs. Altogether, the complete package of specialized cells will form a certain type of organism—whether a giraffe, oak tree, or a human being.

CELL STRUCTURE

Despite their specialization, all cells have the same basic structure, and their protoplasm (the living substance that makes up the cell) consists of two parts: the cytoplasm and the nucleus. The cytoplasm is a jelly-like fluid that contains many tiny, specialized structures, called organelles, each of which performs a particular job. Separate from the cytoplasm—but within the cell body—is the nucleus. This important part directs a cell's activities and contains its genetic "program" that is written and stored in its deoxyribonucleic acid or DNA (the genetic material that carries the code for all living things). It is this program that makes living things different from each other.

Since the cells of both plants and animals share a large number of organelles, there are some similarities. As mentioned, both plant and animal cells have cytoplasm. A jelly-like substance made up mostly of water, the cytoplasm is responsible for keeping the cell alive. Its does this mainly with a division of labor among its mitochondria, endoplasmic reticulum, and ribosomes. Mitochondria "breathe" for the cell, and because they break down food and release energy, they are often described as the "powerhouse" of the cell. The endoplasmic reticulum is a complicated network of tube-shaped membranes that makes and stores a range of substances (like proteins) for the cell. Ribosomes are tiny clusters of granules that actually synthesize or assemble the cell's proteins. Within the cytoplasm of both plant and animal cells are found the common features of Golgi bodies: a vacuole and a cell membrane. The Golgi complex is a collection of membranes that operate as the cell's transport system and store and release various products from the cell. These organelles were named after their discoverer, the Italian scientist, Camillio Golgi (1843–1926). Vacuoles in plants are a large space filled with cell sap that keeps a plant crisp and prevents wilting. In animal cells, vacuoles are smaller and are used for storage and transport. Finally, both plant and animal cells have thin, strong plasma membranes that separate the cell from its surroundings.

Both plant and animal cells also have a nucleus separate from the cytoplasm. The nucleus is the most noticeable feature of a cell since it is large and is surrounded by its own double membrane called the nuclear

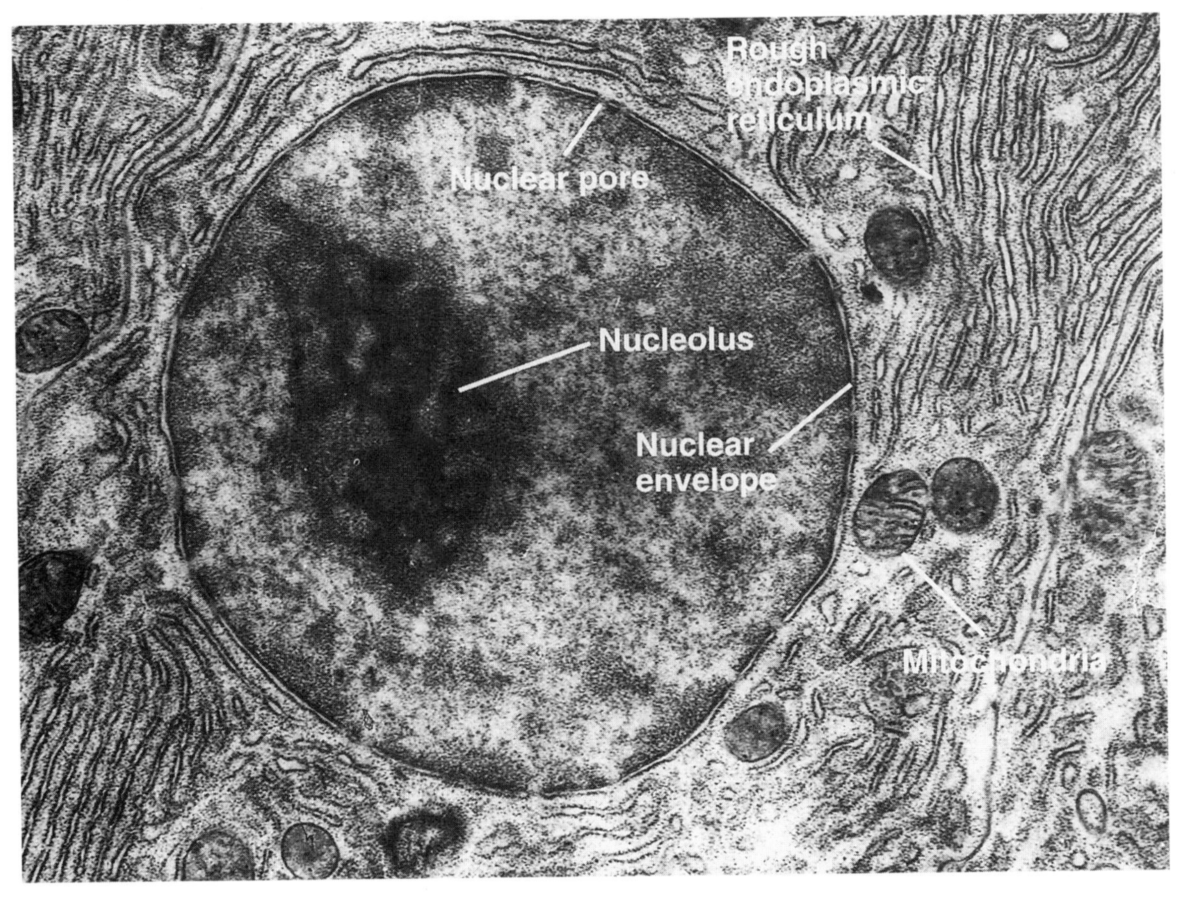

A photograph of animal cells showing some of its common features. All cells have the same basic structure. (©Don W. Fawcett, National Audubon Society Collection/Photo Researchers, Inc. Reproduced by permission.)

envelope. Inside the nucleus are the chromosomes and the nucleolus, which function as the cell's genetic program or chemical instructions; these function mainly when the cell is dividing.

DIFFERENCES BETWEEN ANIMAL AND PLANT CELLS

While plant and animal cells have a great deal in common, each have distinguishing features. Animal cells have lysosomes and centrioles while plant cells do not, yet only plant cells have a cell wall and chloroplast. Lysosomes are small, round bodies that are actually powerful enzymes able to digest or break down the living matter (food) a cell takes in. Centrioles resemble bundles of rods and are important during cell reproduction. Only a plant cell has a cell wall made of a tough carbohydrate called cellulose. Light but strong, it gives the cell its shape and, by fastening together with its neighboring cells, is able to keep a plant growing upright. What most distinguishes a plant cell from that of an animal cell are the existence of chloroplasts and their unique role in the process of photosynthesis. This is the process by which a plant makes its own food in the presence of sunlight. Chloroplasts are disk-shaped sacs that appear green since they contain chlorophyll, which trap the energy in sunlight and produce sugars. Chloroplasts have their own membrane that separates them from the rest of the cell.

SHAPED ACCORDING TO FUNCTION

Cells are shaped differently according to what function they perform in the animal or plant. Some look like snowflakes, others like rods, and some look like round balls. Most are able to move about using hairlike projections called cilia when they are numerous, or flagella when they each have a single tail. Sperm cells use flagella that they whip from side to side to move about. Most cells reproduce by a process called mitosis, in which a cell splits in two and makes an identical copy of itself. Specialized sex cells, which are all genetically different from each other, mix their different genes and produce an entirely different individual cell that has half the genetic content from each of the original two cells.

A SINGLE CELL BEGINS LIFE

The concept of the cell is one of the most important ideas in the life sciences since every living thing begins life as a single cell. Described as a tiny chemical processing plant, it is also the simplest structure able to exist as an individual unit of life. Cells vary in size, shape, and specialization, with the smallest cells belonging to bacteria. The largest cells are the egg cells of mammals and birds. Knowledge of the structure and func-

RUDOLF LUDWIG KARL VIRCHOW

German cell biologist Rudolf Virchow (1821-1902) helped establish cell theory and laid the foundations for modern pathology (the study of diseased tissue). Having demonstrated that the cell theory extended to diseased as well as to healthy tissue, Virchow is considered to be the founder of cellular pathology.

Rudolf Virchow was born in a part of Germany that is now in Poland. After receiving his medical degree from the University of Berlin in 1843, he began working at a Berlin hospital while also teaching at the university. It was during these early years of his career that he first showed his very strong social conscience. While investigating an outbreak of typhus (several forms of infectious diseases caused by bacteria) in Silesia in 1848, he blamed the epidemic on the terrible conditions in which the people lived, thus indirectly blaming the government. Since the ruling establishment was being threatened by real revolutionaries (people promoting political or social change) at this time, Virchow was labeled a radical and lost his jobs. This was not entirely bad for him, since it allowed him time to pursue his scientific research. Thus, by the time he returned to Berlin in 1856 to join a new institute, he had formulated his ideas concerning the importance of cells.

By the 1850s, the cell theory—the idea that all forms of life are made up of cells—had been established for some time, although no one had applied it to pathology. Beginning with his famous statement in Latin, *Omnis cellula e cellula,* meaning that every cell arises from a previously existing cell, Virchow set out to bring the study of disease down past the tissue level and to the cellular level. In 1858, Virchow wrote a book, titled *Cellular Pathology,* in which he demonstrated that the cell theory applied to diseased tis-

tion of cells enables scientists to not only understand better the living organisms they make up, but to treat or prevent things that go wrong at the cellular level. Since the cell is the basic unit of our growth and heredity, a disorder at the cellular level could result in genetic disorders or diseases like cancer in which cells divide in a wild manner, eventually forming tumors. Certain viruses also invade living cells, taking over their machinery to reproduce more viruses. Polio, AIDS, and hepatitis are examples of viruses harmful to humans.

[*See also* **Cell Division; Cell Theory; Cell Wall; Centriole; Chromosome; Cytoplasm; DNA; Gene; Golgi Body; Karyotype; Meiosis; Mitosis; Mitochondria; Nucleus; Nucleic Acid; RNA**]

sue as well. In this work, he was able to show that diseased cells were in fact descended from healthy cells of normal tissue. Although this major work founded cellular pathology and laid the groundwork for later, more fundamental studies of the molecules within the cell, it nonetheless managed to go too far at times. It did this by stating that all diseases happened because of some imbalance or mistake in the cell. In other words, he refused to agree with the germ theory of disease put forth by the French chemist, Louis Pasteur (1822-1895). At times in his career, Virchow's colleagues called him "the pope of medicine," or "the pope of pathology," and it was this stubbornness that led him to refuse to acknowledge Pasteur's research on germs. Virchow's idea of disease has been described as a civil war between cells rather than an invasion of the enemy from the outside. In fact, it is now known that both Pasteur and Virchow were correct depending on the circumstances, since disease can result from an outside invasion as well as from a breakdown of order within the cell.

Perhaps because Virchow was unwilling to compromise on Pasteur's germ theory, he eventually put biology aside and took up anthropology (the study of the origin, distribution, and classification of humans) and archaeology (the study of the material remains of past humans). In fact. because of his major contributions to physical anthropology (the study of the origin and evolution of biological man), he is generally regarded as the founder of that subject as well. Later in his life, Virchow was elected to the Reichstag (the German parliament) and decided to work for social reform and the improvement of public health. He exercised his position's power by designing new sewer systems and hospitals, and worked for better hygiene and food inspection. Still, it is his work in biology that distinguishes Virchow as a major figure in the life sciences. His discoveries can be said to have modernized the entire medical field as well as biology itself.

Cell Division

Cell division is the process by which one cell divides to make two. It is the mechanism that enables an organism to grow, repair damaged tissue, and replace dead cells. There are two different forms of cell division: mitosis and meiosis. Mitosis (my-TOH-sis) is the division of a cell nucleus (a cell's control center) to produce two identical cells. Meiosis (may-OH-sis) is a form of cell division that produces differing sex cells. Mitosis is used to grow, replace, and repair with exact copies. Meiosis is used to produce an entirely new individual.

Without cell division, an organism could no longer grow, reproduce, or repair itself. Every day, the human body makes billions of new cells. Yet each human being began life as a single cell that was formed by the union of a sperm and an egg. Once that single cell began dividing (first into two, then into four, then eight, and so on), the process continued until a complete individual was formed. In organisms that are still growing, like seedlings or children, cells divide very rapidly, but as an organism grows older, many of its cells lose their ability to divide. Thus when cells are dying faster than they can be replaced, the organism begins to feel and show the effects of the aging process, and it looks and acts older. Older people develop wrinkles because their skin and muscle tone is lost as fewer cells are replaced. Older people also cannot heal themselves as quickly as they did when young. All cells have a basic cycle of life that they go through, according to their specialty. A healthy young person's skin cells complete one cycle every twenty-four hours, but a person's brain cells go through only so many cycles and then stop forever.

PHASES OF MITOSIS

Despite the duration of an individual cell's life cycle, each cell goes through the same process when it divides. Most cells are produced by mitosis. In mitosis, a single cell goes through a process in which it eventually produces an identical cell called a "daughter cell." Each daughter cell then grows and soon becomes capable of dividing and producing yet another daughter cell. Mitosis takes place in four stages—prophase, metaphase, anaphase, and telophase—during which each chromosome copies itself, the nucleus divides in two, and the whole cell splits into two identical daughter cells. Each new cell receives a set of chromosomes identical to those of the original cell.

During the first phase of mitosis, called the prophase, the cell's chromosomes become shorter and thicker and duplicate themselves, appearing as double-stranded structures. These joined copies are called chromatids. The membrane around the nucleus also begins to disintegrate. During the metaphase, each pair of joined chromosomes line up across the center of the cell and attach themselves to tiny tubes called spindle fibers. In anaphase, the third stage of mitosis, the chromatids (joined chromosome pairs) are pulled apart by the spindle fibers and move toward opposite ends of the cell as it begins to divide. Actual division occurs in the telophase, when an envelope surrounds each set of chromatids to form a new nucleus in each. Finally, the cell splits in two as the cyotplasm (the cell's jelly-like fluid) turns inward and pinches together, resulting in the production of two new, identical cells. Despite minor differences, mitosis is basically the same for plant and animal cells.

PHASES OF MEIOSIS

Where mitosis makes two identical cells, meiosis produces differing cells. Meiosis takes place whenever reproductive cells such as sperm, pollen, or egg cells are produced. The goal of meiosis is to reduce by half the number of chromosomes, so that when two different reproductive cells join together to form a new organism, it will have the exact same number of chromosomes as its parent. If this halving of chromosomes did not happen, the new cell produced would have twice the number of chromosomes that it should have. For example, in order to be human, an individual must have forty-six chromosomes. Without meiosis, that number would be ninety-two chromosomes after fertilization. Because of meiosis, fertilization will result in an offspring with the exact same number of chromosomes as the parents, getting one-half from each.

Unlike mitosis, meiosis has only two major stages that result in the creation of four reproductive cells. Besides halving the number of chromosomes, meiosis also performs another major function. It allows genetic material to be "shuffled" since chromosomes cross over each other and swap genes before the cell divides. This is a random exchange of genetic

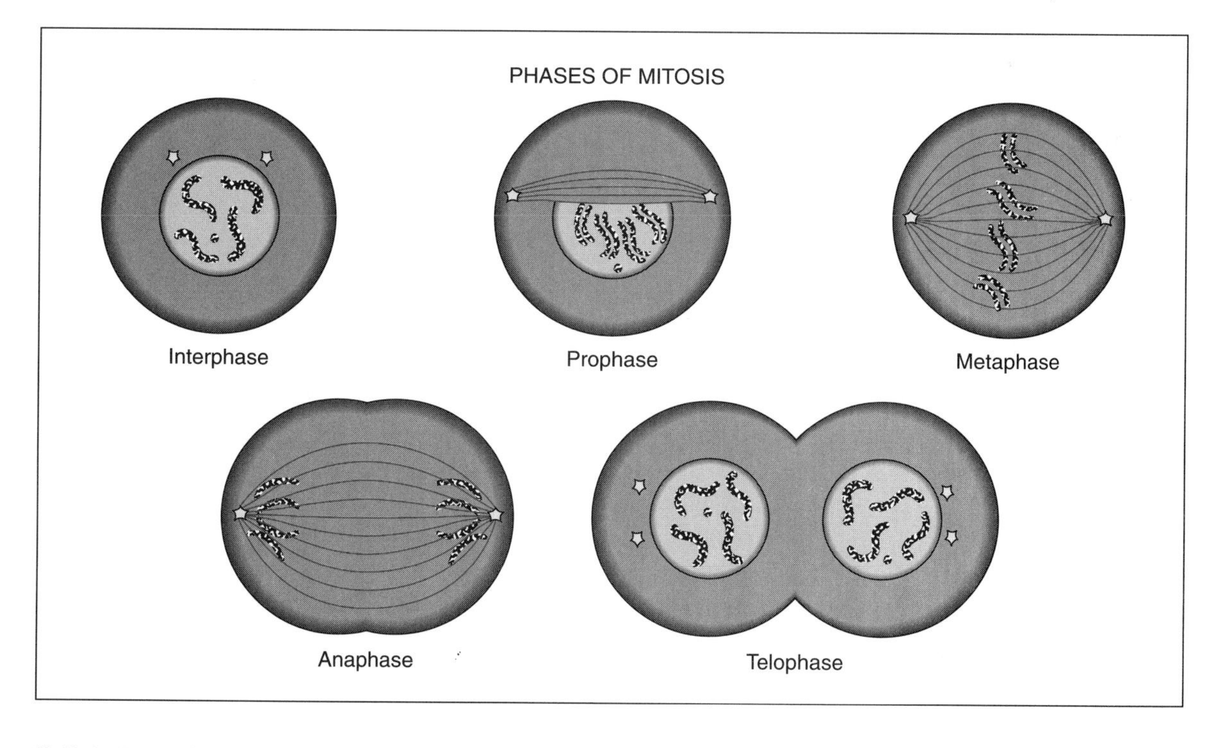

An illustration showing the phases of cell division or mitosis, in which one cell becomes two identical cells. (Illustration courtesy of Gale Research.)

material that assures that an entirely new individual will be produced after fertilization. Because of this shuffle of genetic instructions, each reproductive cell is given its own unique set of instructions (making sure, for example, that no two egg cells will have the exact same combination of genes). This partly explains why brothers or sisters of the same parents have different characteristics. Eventually, when two of these unique cells are joined sexually (sperm and egg) to form a new individual (which further mixes the genetic instructions), a unique organism is created. The exception to this is, of course, the case of identical twins (two complete individuals with the exact same genetic makeup). Identical twins occur after fertilization when a human embryo spontaneously splits during the first cleavage (division) and forms two separate cells.

[*See also* **Cell; Chromosome; DNA; Meiosis; Mitosis**]

Cell Theory

Cell theory states that the cell is the basic building block of all life forms and that all living things, whether plants or animals, consist of one or more cells. It further states that new cells can only be made from existing ones, and that organisms can grow and reproduce because their cells are able to divide.

Although the cell is the basic unit of life, it is also the smallest part of a living organism, and it is therefore too tiny to be seen by the naked eye. Because of this, the notion of a cell did not exist until the seventeenth-century invention of the microscope. Before then, it was commonly believed that the basic units of life were things like fibers and vessels, and that new living things came about by a process simply known as "spontaneous generation" in which life developed suddenly from dead or decomposing matter. By the middle of the seventeenth century, however, some individuals were beginning to investigate the new, subvisible world that the recently invented microscope could bring them. One of these pioneers was the Dutch naturalist Jan Swammerdam (1637–1680), who examined blood and in 1658 became the first person to observe a red blood cell. Also in Holland at the same time, the Dutch naturalist Anton van Leeuwenhoek (1632–1723) developed a powerful, simple (single-lens) microscope and saw, among other things, sperm cells and one-celled animals known as protozoa.

In England at this time, the most famous microscopist was the physicist Robert Hooke (1635–1703), who published his book, *Micrographia,* in 1665. It was in this book that Hooke first used the word "cell" to describe the tiny structures he observed when examining a thin slice of cork under a microscope. He called the structures cells because they resembled a small room (as in "jail cell"). What he was really seeing were the dead remnants of structures that were filled with fluid when the cork was part of a living tree. Although Hooke did not discover living cells, he did coin the word "cell," which was eventually adopted by biologists.

Over the next 150 years, improvements in the microscope allowed observers to better study living tissue, and in 1831, the Scottish botanist Robert Brown (1773–1858) discovered that every plant cell had what he described as a "little nut" or nucleus in it. This discovery paved the way for the cell theory of German botanist Matthias Schleiden (1804–1881) and German physiologist Theodor Schwann (1810–1882). Although others had observed cells and even recognized that animal tissues contained cells, no one had as yet made the connection between cells and life. In 1838 however, Schleiden announced his findings that all vegetable matter is made up of cells and are the fundamental unit of plant life. The following year, Schwann took Schleiden's basic idea and expanded upon it by stating that the cell is the basic unit of *all* living matter, plants and animals. Schwann's clearly stated and well-summarized ideas were more elaborate than Schleiden's and Schwann usually gets most of the credit for establishing cell theory, which he also named. Schwann also suggested that eggs were actually cells and that all life starts as a single cell.

Although not the first to discover cells, Theodor Schwann was the first to clearly state that the cell is the basic unit of all living matter. (Photograph courtesy of The Library of Congress.)

By the middle of the nineteenth century, the final major point was added to cell theory by the German pathologist Rudolph Virchow (1821–1902), who summed up his research with the Latin phrase, *Omnis cellula e cellula,* translated as "all cells arise from cells." Virchow correctly proposed that all cells originate from other cells (putting an end to notions of spontaneous generation), and further demonstrated that even diseased tissue comes from normal cells through the process of division. With this he founded cellular pathology or the study of diseased cells. Cell theory was essentially complete with the 1861 contribution of the Swiss anatomist and physiologist Rudolf Albert von Kolliker (1817–1905), who was the first scientist to study the developing embryo (a living organism in its early stages before birth) in terms of cell theory. Kolliker showed that eggs and sperm should be considered cells and argued that the cell nu-

MATTHIAS JAKOB SCHLEIDEN AND THEODOR SCHWANN

German botanist Matthias Jakob Schleiden (1804–1881) and German physiologist Theodor Schwann (1810–1882) are credited with establishing cell theory as a basic, unifying theme of all biology. The cell theory states that all forms of life are made up of cells and that all living things grow and reproduce because these cells can divide. While it was Schleiden who first formulated the theory in regard to plants, it was Schwann who applied it to both animals and plants and concluded that biology was a single science. Although they were not collaborators, they discussed and compared each other's work, and together offered biology one of its most important concepts.

Matthias Schleiden was born in Hamburg, Germany and began his professional life as a lawyer. He found this field unsatisfying and returned to school to study medicine, later specializing in botany (the study of plants). Using his microscope to study plant tissue, Schleiden eventually concluded that what he was seeing was the most essential unit, or the basic physical unit, of the living plant. In 1838, therefore, he first offered the then-unknown idea that the cell is the fundamental unit of a living plant. As the first to recognize the importance of cells, Schleiden announced that all the various parts of a plant consisted of cells and that they were all created in the same manner. Schleiden was incorrect, however, about a few things, incorrectly stating that all cells developed from the nucleus (a cell's control center), which then disappeared after the cell was fully formed. Yet this does not take away from the overall importance of his work to the world of biology. First, his cell theory of plants focused attention on the basic unit of that organism and second, it laid the foundation for Schwann's broader, more comprehensive work on cell theory.

Theodore Schwann was a very different man than Schleiden. Six years younger, he was as gentle and quiet as Schleiden was impulsive and testy. Where Schleiden would publicly denounce his critics, Schwann would go

cleus was the key to the transmittal of hereditary factors. The only significant modern exceptions added to the original cell theory are that viruses (disease-causing agents) are not composed of cells, although they contain some genetic (hereditary) material and can reproduce in a host cell. Also, mitochondria (produces energy in cells) and chloroplasts (contains chlorophyll to capture the sun's energy) are considered to be parts of cells but contain genetic material and can also reproduce in a cell. As one of the major theories in the life sciences, cell theory serves as the basis for all of today's breakthroughs in the important field of genetics.

out of his way to avoid controversy. Before his work on animal cells, Schwann had done solid work on digestion and was actually the first person to isolate an enzyme (a protein catalyst that speeds up chemical reactions in living things) from animal tissue. An enzyme is a protein catalyst that speeds up chemical reactions in living things. Schwann was able to isolate the enzyme he called pepsin from the lining of the stomach. It was also Schwann who coined the word "metabolism" to describe all of the chemical changes that take place within living tissue.

In 1839, Schwann was working on disproving, yet again, the ancient idea of spontaneous generation (that living things can be generated out of nonliving matter) when he arrived at his own cell theory. Schleiden of course had formulated his cell theory of plants the year before, and in October 1838, the two men got together for dinner and discussed each other's work. Schwann listened to Schleiden's description of what he had seen under his microscope, and said it sounded similar to what he was viewing under his. After dinner, the two went to Schwann's lab and discovered that the cell structure in the spinal cord of a fish was almost the same as that of plants. It was left to Schwann, therefore, to conclude that a cell structure was common to all living things. Having extended Schleiden's theory to animals, Schwann then formulated it in its best and clearest way. The cell theory would prove to unite animal and vegetable biology and show that it was fundamentally one science of biology.

Both men were not entirely correct about everything they proposed, and while their theories did have specific flaws, their work was responsible for formulating one of the most fundamental concepts in biology. The establishment of the cell theory was a landmark achievement in the history of biology. Besides its unifying importance, it also would lead to further research by the best minds who extended the theory and who began to investigate what went on inside the cell.

[*See also* **Cell; Cell Division; Mitosis**]

Cell Wall

A cell wall is a tough, semirigid case that surrounds a cell. Both plants and some single-celled organisms have cell walls. Cell walls are outside the cell membrane and are not part of the living cell. They protect the cell and provide it with support.

Cell walls are found only in some single-celled organisms like fungi and bacteria, but they are found in all plants. They are one of the characteristics that separate plant cells from animal cells (which do not have cell walls). A cell wall is different from a cell membrane, since all cells have plasma membranes that are a part of the living cell. Membranes are also semipermeable and only allow substances of a certain size to pass in and out of the cell. A cell wall in a plant is a structure that is just outside the membrane and provides a plant with protection and rigidity. In plants, it is made up of a complex carbohydrate called cellulose that, although it is very tough, also allows water and solutions to reach the plasma membrane. Since cellulose is both light and strong, it provides the ideal material for a cell wall, acting as a kind of external skeleton that gives the cell (and therefore the plant) its shape and strength. The stem of a plant is able to hold itself up despite gravity by having thousands of cells lined up next to and on top of each other. As the cells take in water, they expand like a balloon and exert pressure against their own walls and against the stem walls. It is their pressure that holds the stem up. When a plant droops, it is because its cells lack water to push against the walls, and the cells begin to shrink.

The cell walls of a green plant are made of cellulose, making it the most abundant organic compound on Earth. The cellulose in a plant's cell walls is formed by fibers that are very strong because they are linked in a criss-cross mesh pattern. Herbivores or animals who eat nothing but green plants must have special digestive systems since the tough cell walls of a plant make it very difficult to digest. This is why herbivores have a much longer and more elaborate digestive tract than do carnivores (meat-eaters) who consume mostly easy-to-digest proteins. Herbivores must also consume enormous amounts of plant material since each mouthful of vegetation contains a relatively small amount of energy (compared to a protein diet).

A plant's cell wall helps protect the important membrane and gives the plant cell and the plant its shape and support. Fungi and bacteria also have cell walls, but they are not made of cellulose. Most fungi have a cell wall made of chitin, while yeast (a type of bacteria) cell walls are made of a different complex of carbohydrates.

[*See also* **Botany; Cell; Plants**]

Centriole

A centriole is a tiny structure found near the nucleus (a cell's control center) of most animal cells that plays an important role during cell division.

Shaped like long, hollow tubes, centrioles help the X-shaped chromatids (duplicated chromosomes) to split apart when a cell divides in two.

All animal cells must divide in order to repair themselves and to grow. Just before a cell divides and produces an identical cell (a process called mitosis), it duplicates or copies its chromosomes so that the new cell will have the same deoxyribonucleic acid (DNA) code as the old cell. Chromosomes contain DNA, which is the chemical that holds the code for all of an individual's inherited traits.

Most of the time, chromosomes are long and thin and appear as a tangled mass of thin threads in the cell nucleus. However, after the chromosomes make an exact copy of themselves and just before cell division is about to take place, the chromosomes begin to shorten and thicken and continuously fold in upon themselves. As they get shorter and thicker, the copy is attached to the original, and together they form a X-shaped structure. Each separate strand of this X-shaped structure is called a chromatid.

While this is going on inside the nucleus of the cell, outside the nucleus small cylindrical tubes called centrioles are preparing to go to work. The centrioles soon move to opposite ends of the nucleus and fibers begin to form between the centrioles as they move away from each other. These fibers make up a structure called a spindle. As division continues to progress, the two connected chromatids are pulled apart by the spin-

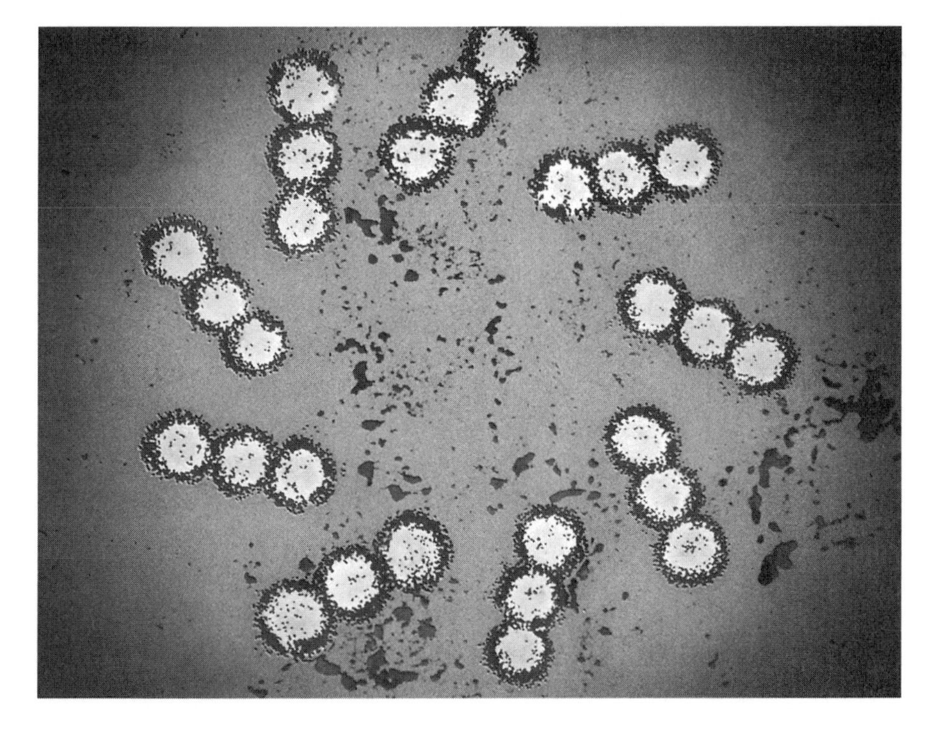

A transmission electron micrograph of cellular centrioles. These tiny structures play an important role in cell division. (©Photographer, Science Source/Photo Researchers. Reproduced by permission.)

dle. Each spindle pulls a chromatid toward it, separating them and splitting each chromosome away from its duplicate. The two sets of chromosomes continue to move away from each other and toward a centriole. As division nears its end, a membrane forms between the two groups of chromosomes and there are now two new identical cells. Although centrioles are still somewhat mysterious to biologists, it is known that they play an important role during cell division since they act as organizing centers for the spindles that actually separate chromosomes.

[*See also* **Cell; Cell Division; Chromatin; Mitosis**]

Cetacean

A cetacean is a mammal that lives entirely in water and breathes air through lungs. Whales, dolphins, and porpoises are cetaceans. Cetaceans are found in all of the oceans of the world, and some live in fresh water.

Whales, dolphins, and porpoises are all members of the order Cetacea. They are mammals that have so completely adapted to life in the sea that they have lost almost all of their hair as well as their hind limbs. Their front limbs have become flat flippers. Their bodies are fishlike and streamlined, with a thick layer of fat beneath the skin that keeps them warm. Despite living in water every day of their lives, they breathe air with typical mammal lungs and therefore must come to the surface to breathe through blowholes at the top of their heads.

All cetaceans possess a certain combination of characteristics. All are endothermic, meaning they are warm-blooded. Unlike cold-blooded animals whose body temperature changes with their surroundings, cetaceans generate their own internal heat and are therefore able to function in most environments. This ability plus their fat layer enables them to exist in near-freezing water.

As mammals, cetaceans bear live young, and their offspring are nourished by milk produced in the female's mammary glands. As water-adapted mammals, cetaceans have entirely lost their rear limbs, and their front ones have changed into fins or flat flippers. They have smooth skin which allows them to move easily through water, and are virtually hairless except when born.

Their body shapes are streamlined, which allows them to move through water with speed and efficiency. They take in oxygen and expel carbon dioxide through a blowhole at the top of their head. This is named after the manner in which they breathe. When a whale breaks the surface after a long dive, it breathes out with such force that the spray can be seen for miles. After inhaling and going under again, the blowhole closes tightly and prevents water from entering the lungs. Cetaceans' respiratory systems have adapted to surviving long periods between surfacings, and they are able to use oxygen in a highly efficient manner.

There are two main types of cetaceans, characterized by what and how they eat. Those that are filter-feeders are called Mysticeti, and those that eat fish and squid are called Odontoceti. Size has nothing to do with what a cetacean eats, since some of the largest whales consume a diet of crustaceans that are sometimes as small as one-tenth of an inch (0.254 centimeters). Filter feeders are also called baleen whales since they have a curtain of fringed plates in the roofs of their mouths (instead of teeth) called a baleen. Using the baleen, these filter feeders strain out tiny shrimp-like water animals called krill. Although baleen whales eat only krill, some, like the blue whale, grow to enormous proportions since they can

Two Atlantic bottle-nosed dolphins. Whales, porpoises, and dolphins are all mammals and members of the order Cetacea. (Reproduced by permission of JLM Visuals.)

eat tons of krill at a time. The great size of the blue whale is possible only because its great weight is mostly supported or held up by water.

Toothed cetaceans are smaller creatures but more numerous. There are eleven species of baleen whales and sixty-seven species of toothed whales, dolphins, and porpoises. Toothed whales eat fish and squid, while dolphins and porpoises eat fish and other sea animals. The main difference between dolphins and porpoises is that dolphins have a beaklike snout and cone-shaped teeth. Porpoises have a rounded snout and flat or spade-shaped teeth.

Cetaceans are considered to be intelligent mammals, having large brains and showing an ability to learn. They are also social animals and usually live in groups. Females care for their young and whales have been known to try to support an ill member of the group and keep it from drowning. All cetaceans have highly developed hearing, since sight is not an important sense for them. Many species use a form of echolocation by making a sound like a click. Using these sounds, they are able to tell how far away something is by how long it takes for their sound to bounce back to them. Some species communicate with one another through clicks, while others, like the humpback whale, use song. Cetaceans reproduce by internal fertilization and usually only a single offspring is produced. Whales have been hunted commercially since the seventeenth century and today many species like the sperm whale are in danger of extinction. Dolphins and porpoises also are killed accidentally by tuna fisherman whose nets trap them below the surface and drown them.

[*See also* **Mammals**]

Chaparral

A chaparral is a geographical region characterized by mild, cool winters and hot, dry summers. Chaparral is sometimes called "scrubland" because the periods of drought and regular fires allow only a certain type of thorny scrub or thicket to thrive there.

A chaparral has what is called a Mediterranean climate because of its mild, moist winters and very dry summers. Besides the Mediterranean however, there are other parts of the world that have a similar climate, such as southwestern Australia, central Chile, the Cape region of South Africa, and southwestern California and northern Baja California, Mexico. The lands in these different areas usually have their own particular names ("mallee" in Australia, "mattoral" in Chile, and "fynbos" in South Africa), and the name chaparral most often refers only to a certain part of California and Mexico. The word chaparral comes from the Spanish word *chapparo* which originally described a thicket of evergreen shrubs.

PLANT LIFE IN A CHAPARRAL

It is the vegetation that grows in this particular climate that most characterizes a chaparral. Typically, a chaparral is composed mainly of woody evergreen shrubs that have adapted to summer drought and fire. There are few tall trees. The leaves of most of these shrubs are small and usually have a waxy, or leathery, covering. Their small size minimizes moisture loss, as does the waxy outer covering. Since the shrubs are evergreen and are not deciduous (they do not lose their leaves all at once), they are ready to absorb rainwater whenever it falls.

Dormancy (in which a plant slows down all its processes) is another way that these plants conserve water loss during a drought. Most chaparral plants have two sets of roots that make them ready to take in any avail-

A chaparral plant community in Diablo State Park in California. This plant life is composed mainly of woody evergreen shrubs that have adapted to summer drought and fire. (©Photograper, The National Audubon Society Collection/ Photo Researchers. Reproduced by permission.)

able water. One of these is a very long taproot that can obtain water deep underground. Another is a lateral root system that grows barely under the surface and can absorb water before it soaks down deeper into the soil.

Fire is another constant in a chaparral, and all of the shrubs that grow there have adapted in some way to surviving fire and sometimes even using fire to their benefit. Chaparral fires happen naturally every ten to forty years, and they are beneficial in the long run. They remove dead plants that built up over the years and release their ash and minerals into the soil to be reused. They also open up the ground, letting more light in and allowing new plants to grow. Most chaparral plants can sprout from their burned base following a fire, and some even need a fire to open their seed coats.

ANIMALS IN A CHAPARRAL

Just like the plants that live in a chaparral, the animals that make their home there have adapted to its extremes. Common chaparral animals in the United States are the mule deer and coyotes. Rodents, reptiles, and rabbits use shrubs to hide from the red-tailed hawk and barn owl. Rattlesnakes and deer mice are also usually abundant. Many of these animals have adapted to their environment since they are able to go without water for long periods, and most avoid activity in the intense midday heat.

Humans are a different story and they are now threatening the natural balance of chaparral. As the chaparral in California becomes an increasingly popular place to build a home, people are having a major effect. Fires that once were natural and needed in a chaparral are suppressed since they would threaten homes. Thus the chaparral becomes thicker and denser, so that when a fire does break out and cannot be contained immediately, the fire that results is ferociously hot and fast-moving, destroying not only homes but even plants that had been fire-adapted. These super-hot fires destroy both shrubs and seeds entirely. Since the seeds or shrubs cannot resprout as before, mud slides often result since there is no vegetation to hold the soil. As a result, these mudslides may cause harm to people, animals, and plants and result in further damage to the chaparral itself.

[*See also* **Biome**]

Chloroplast

Chloroplasts are the energy-converting structures found in the cells of plants. As one of the many tiny organelles (structures inside a plant that have a particular function) in a plant cell, it is the site where photosynthesis (the process by which plants convert the sun's energy into food) occurs. Chloroplasts are not found in animal cells and are the most distinguishing feature of a plant cell.

Chloroplasts allow a plant to capture light energy from the sun and turn it into chemical energy. Chloroplasts accomplish this conversion because they contain chlorophyll, a bright green pigment that absorbs light energy and carries out a chain of chemical reactions. These chemical reactions result in the production of glucose, which the plant uses as food, either storing it or making cellulose to build its cell walls. Chlorophyll is stored in disk-shaped sacs or membranes called thylakoids. It is here that the light energy absorbed by the chlorophyll is directed and changed into chemical energy. This energy allows the plant to take in carbon dioxide, give off oxygen, and eventually produce the plant's food.

A transmission electron micrograph of a chloroplast from a tobacco leaf. Chloroplasts are the organelles the allow a plant to capture light energy from the sun and turn it into chemical energy. (©Dr. Jeremy Burgess/ Science Photo Library, National Audubon Society Collection/ Photo Researchers, Inc. Reproduced by permission.)

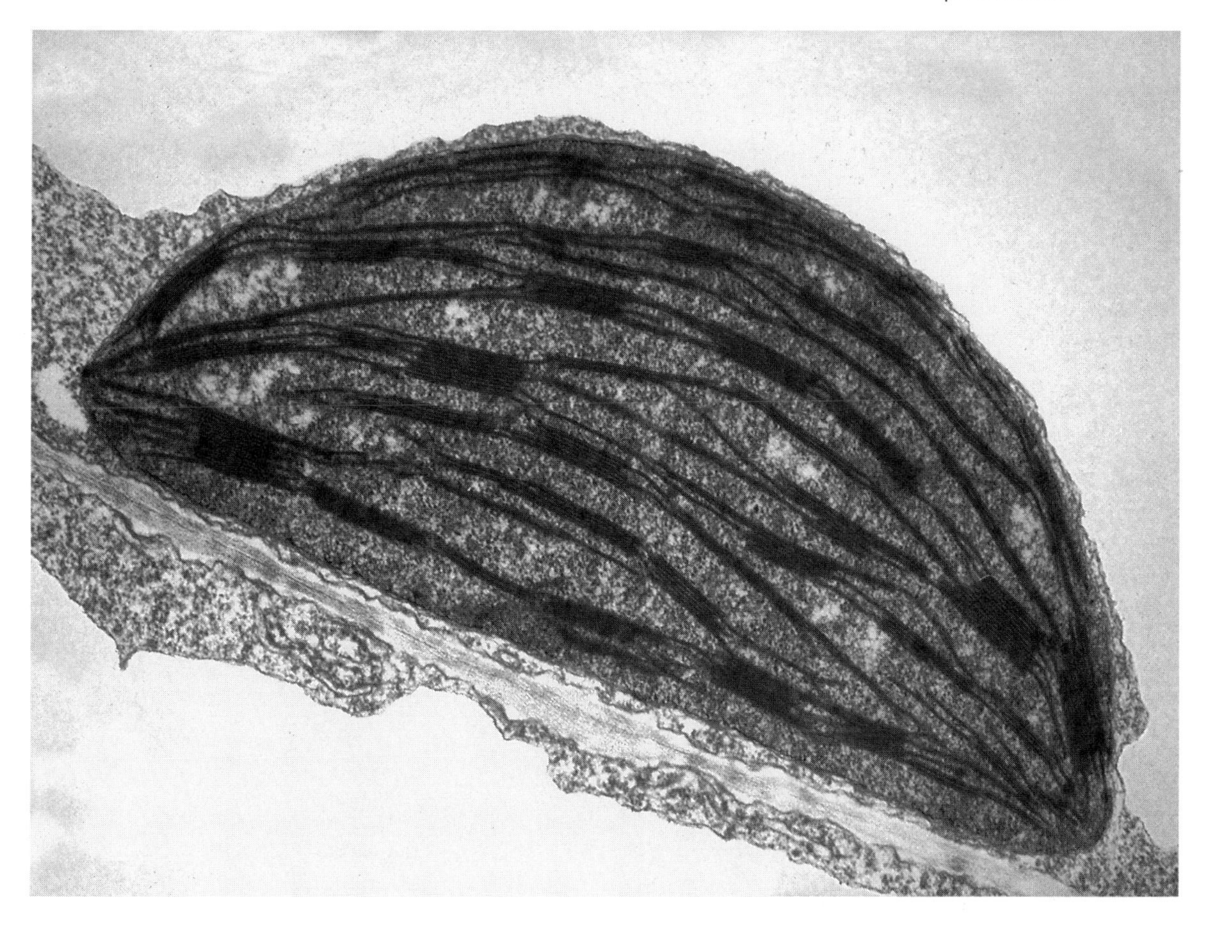

Some plant cells contain only one large chloroplast, but other plant cells may have hundreds of smaller ones. Those areas containing concentrations of chlorophyll are called the grana of the chloroplast, and the spaces between the grana are called the stroma. Inside a plant cell, the chloroplasts are separated from the rest of the cell by a membrane and are usually located around the edges of the cell. Although food production takes place at the cellular level, altogether those cells form an individual leaf on a plant. So it is within the cells of the leaf that photosynthesis occurs, providing the entire plant with food and energy. The energy that the plant has stored is converted back into usable food when the plant is placed in the dark and cannot photosynthesize. Organisms that eat green plants are able to obtain the light energy originally captured by photosynthesis.

[*See also* **Cell; Organelle; Photosynthesis; Plants**]

Chromatin

Chromatin are ropelike fibers containing deoxyribonucleic acid (DNA) and proteins that are found in the cell nucleus and that contract into a chromosome just before cell division. In its unraveled state, chromatin look like beads on a string. In its condensed state, they fold into tight loops that coil up and form x-shaped chromosomes.

When scientists first became able to examine the cell under a microscope using stains to distinguish among its many parts, they noticed that a particular granular material inside the nucleus became more brightly colored by the stain than did other structures. These colored granular structures were named chromatin as derived from the Greek word *khroma* meaning "color." At much higher magnification however, it was discovered that chromatin was not granular but was much more thread-like, with proteins attached to it like beads on a chain. It was soon discovered that in every cell that is not about to actually divide, the cell's genetic material floats about the nucleus as unwound, extremely fine threads or strings called chromatin. In human cells, there are forty-six strands of chromatin forming a tangled mass that has been described as "a bowl of microscopic spaghetti." When a single "noodle" or strand of this mass is examined more closely, it is seen as a coil made up of another compactly folded strand of material which itself is made up a series of loops that are coiled around protein molecules called "histones." It is within these loops that the "twisted ladder" or double helix structure of DNA is found.

Just before a cell is about to divide, this apparently tangled mass of forty-six strands of chromatin begins to condense or gather together to form forty-six easily recognizable, x-shaped packages of genetic information called chromosomes. One of the main purposes of a chromosome is to package the DNA into tight coils so that it all fits into the nucleus. Each chromosome then makes a copy of itself and splits apart, dividing into two identical new cells. In the new cell, the condensed chromosome unravels into its earlier state, containing all the instructions needed to make the cell work and ready to pass on genetic material to the next generation.

[*See also* **Cell; Cell Division; Chromosome; DNA; Nucleus; Mitosis; Protein**]

Chromosome

A chromosome is a coiled structure in the nucleus of a cell that carries the cell's deoxyribonucleic acid (DNA). DNA is the genetic blueprint that contains the genes that both direct the cell's activities and determine the characteristics of the organism. Chromosomes are found in nearly every cell of the body, and different species have different numbers of chromosomes. They are probably the most important part of a living cell since they contain all the necessary information to make a cell work.

Chromosomes were given their name, which means "color body," because they easily take up the dye stain that biologists use to study cell structures under a microscope. Despite this, chromosomes are only visible in the nucleus (a cell's control center) when the cell is dividing (although they are always present). Just before division occurs in a cell, chromosomes are easily seen because they condense or bunch together forming tightly coiled, rodlike shapes (many look like little "X's"). Until this happens, chromosomes exist in the nucleus as unwound, extremely fine threads or strings of protein and DNA that are called chromatin. These loose, strung-out forms of chromatin contain DNA which, in turn, consists of genes. In humans, there are forty-six of these ropelike fibers in the nucleus, which condense or contract into a chromosome just before cell division. Humans have forty-six chromosomes. Chromosomes are bunched together strings of DNA and proteins, and it is these DNA molecules that contain the cellular instructions or coded information that we call genes. The gene is considered the basic unit of heredity. In summary, chromosomes are found in nearly every cell of our bodies. Chromosomes are made of DNA, and DNA stores genes. It is genes that carry the vital codes and information that not only tell a cell what to do, but which get passed on to the next generation by sexual reproduction.

THOMAS HUNT MORGAN

American geneticist (a person specializing in the study of genes) Thomas Hunt Morgan's work with the fruit fly established the chromosome (a coiled structure in the nucleus of a cell that carries the cell's deoxyribonucleic acid, or DNA) theory of inheritance. He discovered that chromosomes are composed of discrete units called genes which are the actual carriers of specific traits. He also showed that when genes mutate, or change, the traits they control also change.

Thomas Hunt Morgan (1866–1945) was born in Lexington, Kentucky and grew up surrounded by nature and wildlife. As a youngster, he had an intense interest in biology and later majored in zoology in college. After obtaining his Ph.D. from Johns Hopkins University in 1890, he taught for awhile, and in 1904 became professor of experimental zoology at Columbia University where he would remain until 1928. Morgan had long been interested in heredity and believed that it was an important phenomenon barely understood. When Morgan began his research in 1904, the world had just learned of Austrian monk Gregor Mendel's laws of heredity. One of these laws stated that mixing traits did not result in a blend of traits, but instead these traits sorted themselves out according to a fixed ratio. Most scientists realized that the behavior of newly discovered chromosomes during cell division seemed to match Mendel's laws. However, everyone knew that there were only slightly more than two dozen pairs of chromosomes in the human cell, and it did not seem possible that they alone could account for the huge range of inherited characteristics exhibited by people. The explanation might be that each chromosome contained large numbers of different "factors," or "genes."

In 1907, Morgan decided to attack this problem using a new tool to science called the *Drosophila melanogaster,* or the common fruit fly. Fruit flies are the very tiny flies attracted by the smell of fruit. Leave a bowl of fruit out in the summer, and chances are these flies will somehow get in the house

Genes have been compared to instructions or to a recipe for making proteins. Proteins can be found in virtually every part of the body, and they help cells do all the complicated things they have to do. There are somewhere between 50,000 and 80,000 genes in the human body, and each contains instructions on making a protein that has a specific purpose. There are genes for proteins that make our eyes, our organs, our hair, and our skin. There are genes that influence how tall we will be or what our skin color is. One of the main functions of chromosomes is to package the DNA that contains these genes that tell our cells what type

and swarm onto the peaches and plums, attracted by the odor. For Morgan, however, these flies proved perfect for genetic research since they were inexpensive, very easily bred in large numbers, multiplied rapidly, and best of all, their cells possessed only four chromosomes. Morgan would therefore tackle the problem of inheritance by closely following the generations of flies. It was in his famous "Fly room" with his undergraduate students Calvin Blackman Bridges (1889–1938), Alfred Henry Sturtevant (1891–1970), and Hermann Joseph Muller (1890–1967), all of whom went on to do major work in genetics, that Morgan discovered many instances of mutations. He was able to trace these in later generations and to prove that genes were linked in a series on chromosomes (or inherited together) and were responsible for identifiable, hereditary traits. Morgan was also the first to explain sex-linked inheritance when he located the mutant white-eye gene on the male sex chromosome (fruit flies should all have red eyes). He further explained the "mistake" phenomenon called crossing-over in which traits found on the same chromosome are not always inherited together. In this, Morgan showed that one chromosome actually exchanged material with (or crossed over to) another chromosome. This mistake proved to be an important source for genetic diversity since it can possibly add unpredictable variety.

By 1911, Morgan and his "Fly room" team had created the first chromosome maps for fruit flies. In 1915, Morgan and his students published a summary of their work, *The Mechanism of Mendelian Heredity*, which would lay the groundwork for all future genetically based research. Morgan's rigorous work with the humble fruit fly enabled him to discover how genes are transmitted through the action of chromosomes, thus confirming Mendel's laws of heredity and laying the foundation for modern experimental genetics. In 1933, Morgan was awarded the Nobel Prize in Physiology or Medicine for his work on heredity. His student, Hermann Muller, also experimented with fruit flies and proved that x rays can damage genetic material. For this, Muller received the 1946 Nobel Prize.

of proteins to make. This protein-making is a nearly constant activity in the cell and can be considered a kind of biological housekeeping.

MITOSIS

Every day our bodies make billions of new cells that are identical to the ones they will replace. This is because every cell in our bodies has its own life cycle, and some, like skin cells, complete their full cycle in only twenty-four hours. In this steady production of identical or "sister" cells, called mitosis, each chromosome makes a copy of itself, moves to the opposite ends of the cell membrane, and splits into two identical cells after a membrane forms across the cell's middle. This process assures that each new cell gets the correct genetic material.

MEIOSIS

During sexual reproduction, however, an entirely different process called meiosis takes place that involves chromosomes. When a sperm fertilizes an egg, each sex cell (sperm and egg) starts with only twenty-three individual chromosomes, unlike all other cells in the body that have a full set of forty-six (or twenty-three pairs). When the sperm and egg join together, the first new cell created gets twenty-three chromosomes from the mother (egg cell) and twenty-three chromosomes from the father (sperm cell), to form a full complement of forty-six chromosomes. Meiosis also adds a final shuffling of genes that happens before division takes place. During this shuffle, chromosomes cross over each other and actually swap genes, thus further assuring that each sex cell has its own unique combination of genetic instructions. Unlike mitosis, the new cell (and eventually new organism) created is not identical to the cells that formed it but is rather a mixture of the chromosomes of two organisms. This is why in

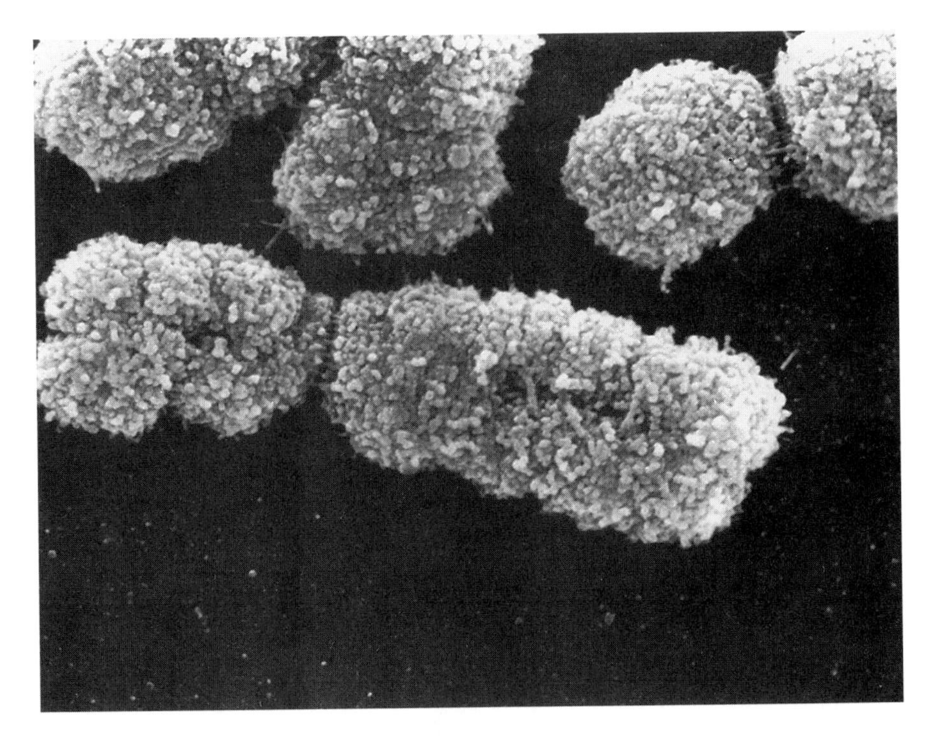

A scanning electron micrograph of a human X chromosome. Chromosomes are probably the most important part of a living cell since they contain all the necessary information to make a cell work. (©Biophoto Associates, National Audubon Society Collection/Photo Researchers, Inc. Reproduced by permission.)

organisms that reproduce sexually, an offspring does not look exactly like either parent since it inherited genes from both.

At the chromosome level, the difference between a male and a female is only one gene on one chromosome. Chromosomes are always in pairs, and those that determine the sex of an individual make up two of our forty-six chromosomes. These two are known as sex chromosomes. The other forty-four chromosomes are not involved in determining sex and are called autosomes. Females have a pair of sex chromosomes called XX, while males have a pair called XY. Thus the difference is only one (Y) chromosome. If a human embryo is given two X chromosomes, a certain area of cells becomes the egg-making part (the ovaries) and the embryo will develop into a female. If it has an X and a Y chromosome, the Y signals the cells to start producing sperm-making parts (testes), and the offspring develops into a male.

Nature, however, can and does make mistakes (usually during meiosis), and when they occur at the chromosome level, they can be disastrous. The most common mistake in humans is called aneuploidy. This occurs when an offspring has an extra or a missing chromosome. Most cases of aneuploidy result in the mother aborting her fetus spontaneously (called a miscarriage). This can be considered nature's way of putting an end to a mistake. One instance in which fetuses do develop and are born is that of Down's syndrome in which the offspring has an extra chromosome. However, these people suffer from mental retardation and some physical deformities.

[*See also* **Cell; DNA; Genetic Disorders; Genetic Engineering; Genetics; Mendelian Laws of Inheritance; Mutation; Nucleic Acid; Protein**]

Cilia

Cilia are microscopic, hairlike structures that project from the edges of certain types of cells (the building blocks of all living things) and allow them to move themselves or things that are close by. Not all cells have cilia, and those that do are usually animal cells rather than plant cells. In higher animals, such as humans, cilia also refer to the hairlike lining of the nose, ear, and trachea (the air passage to the lungs) that keep those passages clean from dust, pollen, bacteria, and mucus.

Animal cells must often move about, and cilia are the primary means by which they achieve movement. Cilia are composed of microtubules or extremely tiny tubes whose action or movement can be controlled by the cell. By coordinating the wavelike action of its cilia, the cell can either send itself through its environment or help to move the environment past itself. If one-celled organisms like the protozoan *Paramecium* are observed under a microscope, the rhythmic, wavelike motion of its cilia can be easily seen beating against its liquid environment as if it were rowing in a coordinated way. Certain one-celled organisms also use their cilia to capture food and move it into their gullet.

Certain cells, like gametes or sex cells, only have a single projection that they use to move about. When a cell has this sort of singular, long, hairlike projection that resembles a tail, it is called a flagellum instead of a cilium. Sperm cells are a good example of cells that have flagella.

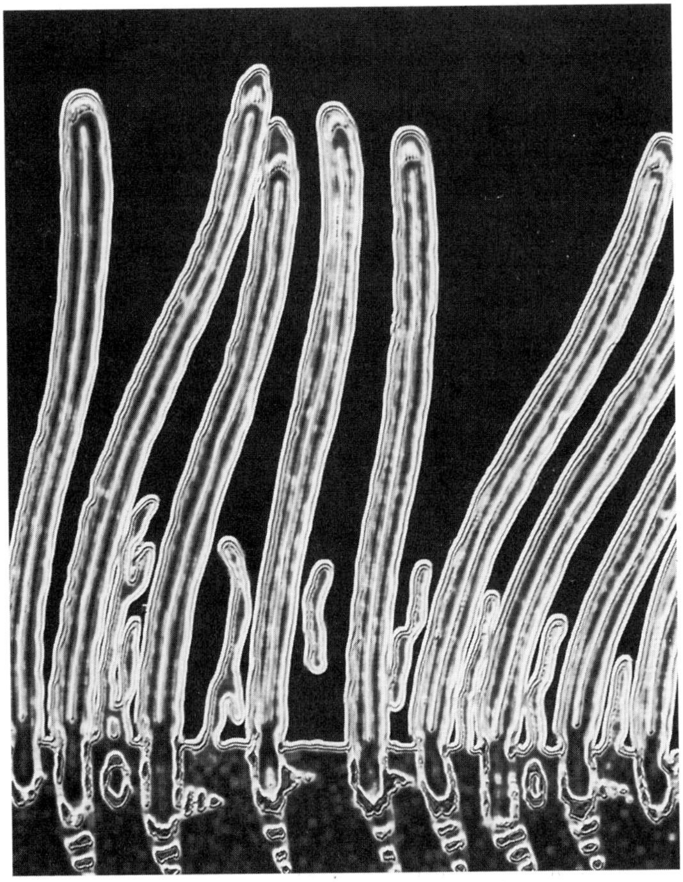

A computer graphic of cilia found on the surface of the human windpipe, or trachea. (©Photographer, Science Source/Photo Researchers. Reproduced by permission.)

Besides the epithelial (skin) tissues in most higher animals contain a carpet of cilia whose purpose is to move tiny particles across and away from sensitive surfaces. The wavy motion of the cilia in the uterine tubes in women help move and guide the fertilized ovum (human egg) down to the uterus where it can attach itself and grow into a fetus. Even a clam uses its cilia to fan water containing oxygen into its gills. Cilia are so tiny that just the trachea may contain as many as 1,000,000,000 cilia per square centimeter.

Circulatory System

The circulatory system is a network that carries blood throughout an animal's body. Described as an internal transport system or a distribution system, the circulatory system maintains a constant flow of blood throughout the body, carrying nutrients and oxygen to the body's tissues and taking away its waste products. It also helps regulate the body's temperature; carries substances, like antibodies and white cells that protect the body from disease; and transports chemicals, such as hormones, that help the body regulate its activities.

All animal cells live in a liquid environment, taking in oxygen and nutrients and creating waste products. All, therefore, require a way to gain access to what they need to grow, to perform their specialized task, to reproduce, and to dispose of their waste products. For any animal larger than a single-celled organism—which meets its needs by passing substances through holes in its membrane (called diffusion)—a complex circulatory system is required in order to make sure that every single cell gets what it needs.

OPEN CIRCULATORY SYSTEM

There are two main types of circulatory systems—an open system and a closed system. Most invertebrates (animals without a backbone) have an open circulatory system. It consists of a fairly simple network of tubes and hollow spaces. In this system, a heart (or a series of hearts) pumps blood out of the vessels (a duct for circulating blood) and into the sinuses or open spaces of an animal's body. Mollusks, like clams, and arthropods, such as crayfish and grasshoppers, have an open circulatory system. In such a system blood flows slowly and under low pressure into the open spaces in these organisms' body cavities. This blood also bathes their cells, allowing them to obtain food and oxygen while eliminating waste. Open circulatory systems pump a bloodlike fluid called "hemolymph" that closely resembles seawater. However, one of the disadvantages of this type of system is that it cannot respond quickly to change, nor can it supply large amounts of oxygen.

CLOSED CIRCULATORY SYSTEM

All vertebrates (animals with a backbone) and some invertebrates, like earthworms, have a closed circulatory system. This means that the blood never leaves their vessels. For vertebrates, blood is pumped by the heart throughout the body via a network of closed vessels that become finer and finer as they get farther away from the heart. Closed systems are more efficient than open ones, since it is easier for them to respond to sudden changes and to alter the distribution of blood.

Although there are invertebrates with closed systems, it is by far the predominant characteristic of vertebrates—whether fish, amphibian, reptile, bird, or mammal. The basic components of the vertebrate circulatory system are the heart, arteries, capillaries, veins, and the blood itself. In the human body, the heart is a hollow, muscular pump that forces blood to move throughout the body. The human heart consists of two pumps that lie side by side. The left and stronger pump receives oxygen-fresh blood from the lungs and pumps it under great pressure to the cells throughout the body. The weaker right side receives "used" blood from the cells and sends it to the lungs to have carbon dioxide removed and to be freshened with oxygen. The heart muscle beats at its own automatic rhythm and pumps in a certain correct sequence. Arteries are blood vessels that carry blood away from the heart. They are specially designed, with thick, strong walls, since the blood they transport is under high pressure. Puncturing an artery can cause blood to spurt and travel through the air. Arteries also serve to smooth out the flow of blood by absorbing much of the rhythmic shock of the pumping heart. Arteries get smaller farther from the heart and are called arterioles.

As oxygen-rich blood continues the one-way trip from the heart and to the cells, the capillaries receive blood from the arterioles and pass it directly to the surrounding tissues and cells. Capillaries are finer than human hair (they are one cell thick) and have very thin walls where the critical exchange of nutrients, oxygen, and waste takes place. Capillaries merge to form venules or tiny veins, which in turn, merge to create large veins that will carry the blood back to the heart. Large veins are thinner than arteries (since the blood is now under much lower pressure) and are the primary blood vessels for the one-way, return trip of blood to the heart. Large veins also have one-way valves that prevent deoxygenated blood (depleted of oxygen and full of carbon dioxide) from flowing backwards. These veins empty into the parts of the heart called the vena cava, which in turn, empty directly into the heart's right atrium and back on to the lungs to start the cycle all over.

The route that blood travels in humans (as well as in birds and other mammals) is called "double circulation," which is in contrast to animals whose blood circulates in a single loop from the heart, around the body, and back again. In a double circulation system blood flows alternately through the lungs and throughout the body in a figure-eight pattern. After completing each loop, blood returns to a different side of the heart. In the human body, an entire double-loop cycle takes less than one minute. While the actual path that blood follows will vary among different groups of vertebrates, all are based on a similar system whether single or double loop. For example, fish have single loop circulation and a four-chambered heart, while amphibians and most reptiles have a three-chambered heart.

Opposite: An image of the main components of the human circulatory system. The heart (placed between the lungs) delivers blood to the lungs, where it picks up oxygen and circulates it throughout the body by means of blood vessels. (Drawing courtesy of Gale Research.)

WILLIAM HARVEY DISCOVERS BLOOD CIRCULATION

The English physician William Harvey (1578–1657) discovered blood circulation in 1628. He demonstrated, contrary to widely-held beliefs that

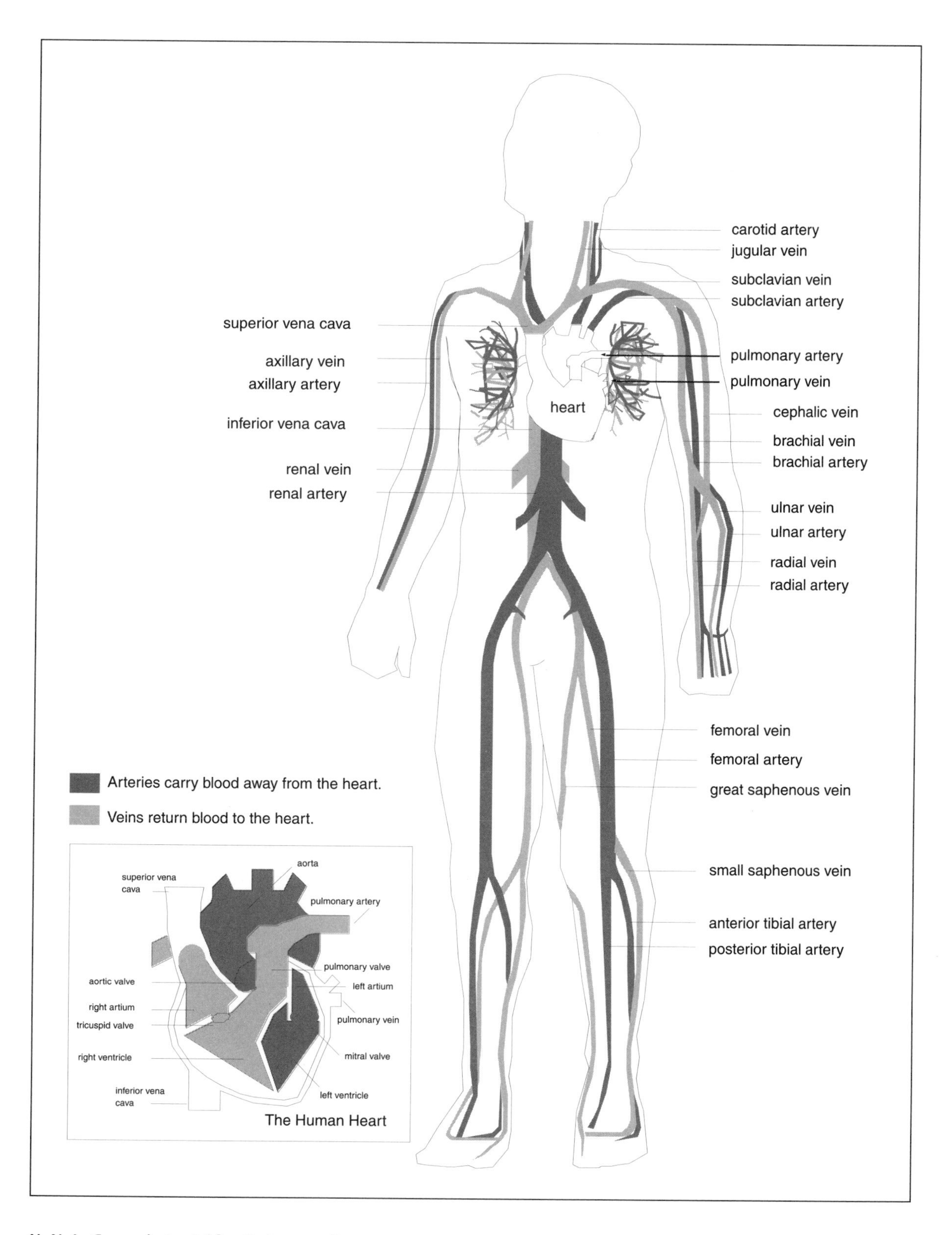
carotid artery
jugular vein
subclavian vein
subclavian artery
superior vena cava
axillary vein
axillary artery
inferior vena cava
renal vein
renal artery
pulmonary artery
pulmonary vein
heart
cephalic vein
brachial vein
brachial artery
ulnar vein
ulnar artery
radial vein
radial artery
femoral vein
femoral artery
great saphenous vein
small saphenous vein
anterior tibial artery
posterior tibial artery
Arteries carry blood away from the heart.
Veins return blood to the heart.
superior vena cava
aorta
pulmonary artery
pulmonary valve
left artium
aortic valve
right artium
tricuspid valve
pulmonary vein
right ventricle
mitral valve
inferior vena cava
left ventricle
The Human Heart

WILLIAM HARVEY

English physician and biologist William Harvey (1578–1657) was the first to describe the circulation of the blood through the heart and the blood vessels. As a great experimenter, Harvey was able to prove that blood does not ebb and flow like waves in the body as people believed, but instead travels in a closed, or one-way circle. Many consider Harvey to be the founder of modern physiology (which studies how the different processes in living things work).

William Harvey was born in Folkestone, England, the son of well-to-do parents. After earning his degree from Cambridge in 1597, he did what anyone who wanted the best medical education available did: he went to Padua, Italy. At that time, Padua was able to boast that it had many great scholars, including the young astronomer, Galileo Galilei (1564–1642), on its faculty. Harvey stayed in Padua until he obtained his degree in 1602, and then he moved back to England. Given his father's connections, he soon had important patients, and later became physician to the king. However, Harvey was more interested in experimentation than in his day-to-day medical practice, as evidenced by his claim to have dissected eighty different species of animals by 1616. Harvey always dissected with a purpose, and most often, he focused on learning more about the heart and its vessels. Because of his extensive hands-on experience, Harvey already knew that heart muscle was basically a pump that acted by contracting, which pushed the blood out. He also learned that the valves that separated the heart's two upper and two lower chambers were one-way valves, and that veins (a branching system of vessels through which blood returns to the heart) even had one-way valves. All of this was leading Harvey far away from what most medical men of his time believed.

Traditional medicine was based on the writings of Galen (A.D. 129–C.199) who lived some 1,400 years before Harvey and whose human physiology was based on his dissections of animals. Galen taught that the liver was the im-

blood flowed from the heart in a continuous, one-way cycle. The current understanding of the circulatory system was born of Harvey's work. It is now known that the circulatory system performs many vital functions in respiration (delivering oxygen to the cells and removing carbon dioxide); in nutrition (carrying nutrients to the cells and liver); and in waste removal (transporting poisons like salts and ammonia to the liver for disposal). The circulatory system also helps the body fight disease by acting as a means of transport for the lymphatic system, which is part of the body's immune system. This system filters harmful substances out of the

portant organ for the blood and that the blood that formed there ebbed and flowed throughout the body like the back-and-forth of the oceans' tides. Before Harvey decided to refute Galen, who he knew to be wrong, he performed even more experiments on animal and human hearts. He monitored the beating heart and made rough calculations of the volume of blood leaving the heart per beat and then compared that to total blood volume. It seemed impossible to him that blood could be broken down and reformed, as Galen said, fast enough to account for the amount of blood in the human body. The only explanation was that it must be the same blood moving in circles throughout the body. Finally, in 1628, Harvey published his slim, seventy-two-page book in Holland, in which he argued that blood was in motion all the time and that it circulated. The translated title of his landmark work is *On the Motions of the Heart and Blood.*

Since Harvey's book was a direct assault on Galen and all those who supported his ancient ideas, it is not surprising that Harvey found himself being ridiculed and condemned. His medical practice suffered for a time, but Harvey refused to be drawn into a defensive debate and instead, let the facts speak for themselves. Throughout this, he retained the king's favor. By his old age, the theory of the circulation of the blood was accepted and he had become a highly respected and even revered man of science. Interestingly, the one part of Harvey's theory that he could not prove was the crucial statement that blood moved from the arteries (a branching system of blood vessels that carry blood away from the heart) to the veins. He was unable to point out any visible connections between the two. However, he knew from his dissections that both vessels divided down to smaller and finer vessels, again and again, and so he simply assumed that they were too small to see with the naked eye. He was, of course, correct and was proven so about twenty-five years later when his countryman, Marcello Malpighi (1628–1694), used his new microscope and discovered extra-fine blood vessels called capillaries that connected the smallest arteries to the smallest veins.

bloodstream and carries white blood cells that destroy harmful bacteria, viruses, and other invaders.

[*See also* **Blood; Heart; Lymphatic System; Respiratory System**]

Class

The term class refers to one of the seven major classification groups that biologists use to identify and categorize living things. These seven groups are hierarchical and range in order of size. Class is the third largest group and is located between phylum and order. The classification scheme for all living things is: kingdom, phylum, class, order, family, genus, and species.

Members of a class have more characteristics in common than do members of a phylum. The following example compares animals that are included in phylum with those in class. In the former are included mammals, reptiles, and birds, yet each of these are placed in a different class. Humans and mice belong to the class Mammalia because they are characterized by having hair or fur on their bodies and feeding milk to their young. Lizards and snakes belong to the class Reptilia since they are covered with scales and do not feed milk to their young. Sparrows and eagles belong to the class Aves because they have feathers on their bodies and do not feed milk to their young. All, however, belong to the same phylum Chordata since they all have a vertebrae or backbone.

In addition to these obvious differences between classes, such as body covering and the way the young are nurtured, there are more subtle ways of determining an organism's correct class. One of these is called comparative biochemistry. This process compares the deoxyribonucleic acid (DNA) of different organisms to determine if their genetic information is the same. Since the grouping by class is near the midpoint of the seven groupings, the organisms that will remain and be lumped in the next group, order, will have even more in common with each other.

[*See also* **Classification; Family; Genus; Kingdom; Order; Phylum; Species**]

Classification

Classification is a method of organizing plants and animals into categories based on their appearance and the natural relationships between them. Also called scientific classification, it is science's way of identifying and grouping living things. The classification of organisms is a science called taxonomy, or systematics.

The first person to attempt any type of systematic grouping of organisms was the Greek philosopher and scientist, Aristotle (384–322 B.C). Until his work, most people simply divided all plants and animals into two basic categories: useful and harmful. As a careful observer of the natural world, Aristotle began arranging organisms according to their physical similarities. Since there were only about one thousand organisms known in his time, he classified animals according to those with red blood (vertebrates or having a backbone) and those with no red blood (invertebrates or no backbone). He also classified plants by size and by whether they were herbs, shrubs, or trees. Despite many mistakes and an oversimplified idea, Aristotle's impulse to classify and to categorize organisms was a necessary attempt to make sense of the diversity of life in order to study it better.

THE BENEFITS OF CLASSIFICATION

In the life sciences, the need to organize is very important and extremely useful. Classification helps biologists keep track of living things and to study their differences and similarities. It also shows biologists how living things are related to one another through evolution (the process by which living things change over generations). Classifying also saves time and effort. There are many possible ways to classify life: appearance, behavior, evolutionary history, or life development from fertilization to adulthood. The modern classification system is considered a natural system since it represents genuine relationships between organisms. In this natural system, the more closely organisms are related to each other, the more features they have in common. This system is also hierarchical, meaning that its categories are grouped according to size in a series of successively larger ranks.

CAROLUS LINNAEUS DEVELOPS BINOMIAL SYSTEM OF NOMENCLATURE

The system used today is based on the work of one individual, the Swedish physician and naturalist, Carolus Linnaeus (1707–1778). In his day, it sometimes took as many as ten words to name a particular organism and no standard system existed upon which everyone agreed. Linnaeus traveled throughout Europe compiling lists of the animals and plants he encountered, and in 1735 published a book which tried to make some sense of this great diversity. By 1758 he had completed his huge encyclopedia called *System of Nature* which described and classified all known organisms by their structure and placed them in one of the seven levels of his hierarchical system. Linnaeus also developed the binomial system of nomenclature, which gave a distinctive two-word name to each species. This system is still followed with the first part being the genus name, and the second part serving as its species name. For example, both the bobcat and the house cat belong to the genus *Felis,* but the bobcat's species is *rufa* (*Felis rufa*) while that of the domestic cat is *Felis domestica.* The second name is usually descriptive of the particular animal. Among the rules for this system, the two-part name is always used. The species part

CAROLUS LINNAEUS

Swedish botanist (a person specializing in the study of plants) Carolus Linnaeus (1707–1778) devised the first orderly system of classifying living things. He also introduced the binomial system of nomenclature (a two-part naming system) that is still in use today. He is called the father of taxonomy (the science of classifying living things) because his system was able to impose a much-needed order on the study of life itself.

Being able to identify a plant or animal, to tell how it is different from others, and to know how it fits into the entire natural world is something that people simply take for granted in today's world. However, there was a time in the history of the life sciences when naturalists (people specializing in the study of plants and animals in their natural surroundings) used as many as ten words to give something a specific, descriptive name, and even with all that effort, there was no guarantee that it would be used by others or that someone who spoke a different language would know what the name meant. This describes what the life sciences were like before the great classifier, Carolus Linnaeus, gave science a practical way of naming organisms that was based on clear and simple standards upon which everyone could agree.

Linnaeus was born Carl von Linne in South Rashult, Sweden. (Linnaeus is the Latin version of his name.) His father was a clergyman, and the very young Linnaeus was so interested in gardens and growing things that the locals called him "the little botanist." When his father sent him to medical school, Linnaeus was able to combine school with botanical exploring trips that only made him more interested in plants. When he became lecturer in botany at Uppsala University at the age of twenty-three, he was able to go on longer, more extensive excursions to Lapland in 1732. After traveling 4,600 miles (7,407 kilometers) throughout northern Scandinavia discovering new plant species and observing animal life, he began to formulate the details of an idea that he had first expressed in a paper some years before. By 1735, Linnaeus had published his *System of Nature* in which he proposed the idea of classifying plants in the simplest and most clear way that one could. To Linnaeus, that meant a system based on the specimen's external characteristics that were most obvious to the eye. His system would there-

is never used alone. The generic name always begins with a capital letter but the species name is always lowercase. Both names are written in *italics* or are underlined. Latin is used to avoid any confusion in translating different languages. Altogether, this system allows everyone in the world to use the same name for the same organism and to immediately understand each other.

fore be based on observable characteristics such as structure, or anatomy, or on the details of the way a thing reproduced itself. In his landmark book, he showed how this could be done. First, he created a hierarchical system (in a hierarchy, things are arranged in a certain order) in which the categories above included all of the ones below it. Thus, he created a system in which living things were grouped according to their similarities, with each succeeding level possessing a larger number of shared traits. He named these levels *class, order, genus,* and *species.* He also popularized what is called binomial nomenclature, which gave every living thing a Latin name consisting of its genus and species. For example, this would distinguish two very different species, like a lion and a cougar, simply by their Latin names. The lion belongs to *Panthera leo* and the cougar belongs to *Felis concolor.* Thus each organism has a generic name, telling which group it belongs to, and a specific name for itself.

Although classification might not seem to be as important a subject to science as some others, it proved absolutely essential to such a broad and diverse field as the life sciences. In fact, only after Linnaeus's system was accepted and regularly used did biology and botany begin to make real progress. The advantages of his system are numerous. For example, his use of Latin allows scientists to communicate worldwide about organisms without having to understand different languages. Since each type of organism can fit into his system in a logical and orderly way, it can be expanded indefinitely. It is also a great advantage that the levels of his hierarchical system provide a framework for seeing and understanding the relationships among different organisms or groups of organisms. His system is also flexible and adaptable. Since it was first introduced, the number of levels have grown, with *phylum* being inserted above *class* and *kingdom* being placed at the very top. Finally, although his system was introduced before the theory of evolution (the process by which gradual genetic change occurs over time to a group of living things), it always has been able to accommodate any new discoveries or modifications which that theory has made. Linnaeus was said to have been an almost obsessive classifier, yet he was a person who turned his passion for an idea into a truly great scientific contribution.

CLASSIFICATION GROUPS

Seven major groups or categories make up the scientific classification system. The groups or categories themselves are called taxons (from taxonomy, which is the science of naming and classifying organisms). These groups range in order of size, so from the largest or most general to the smallest and most specific, they are: kingdom, phylum, class, order, family, genus, and species. Each kingdom is divided into smaller and smaller groups until each type of organism is placed in a unique category. One way of remembering this general-to-specific scheme is the rhyme or formula, "*K*ing *P*hilip *C*ame *O*ver *F*rom *G*reat *S*pain."

Kingdom is the largest unit and is composed of five separate kingdoms: Monera, Protista, Fungi, Plantae, and Animalia. Beginning with Linnaeus and for a long time afterward, there were only two kingdoms, Plantae and Animalia. But with the improvement of the microscope and the discovery of microorganisms, the number was expanded to five. From kingdom on down to species, organisms are grouped together with increasing similarity. Besides these seven major groups, biologists are able to use various subgroups to deal with minor differences among organisms when those differences are not large enough to form a new group. For example, species may be divided up into subspecies.

Classifying a dog and a wolf offers a good example of how two animals would fit into these seven categories. Both are in the kingdom Animalia since they cannot make their own food. Next, they would both be in the phylum Chordata since they have a notochord (like a vertebrae or backbone). Both also belong to the class Mammalia since they have fur and feed milk to their young. Both are members of the order carnivora since they are meat eaters. They also both belong to the family canidae because they cannot retract their claws and they hunt and stalk their prey. However, while both are similar enough to be in the same genus, canis, they are different enough to be in separate species. Therefore, the wolf's scientific name is *Canis lupus* and the dog's is *Canis familiaris.*

A classification system provides a method that best represents genuine relationships between organisms. It is a natural system that is based on overall resemblances and which reflects how each organism is related from an evolutionary standpoint.

[*See also* **Class; Family; Genus; Kingdom; Order; Phylum; Species**]

Cloning

A clone is a group of genetically identical cells descended from a single common ancestor. A clone can describe a group of cells or a multicellular organism. In both cases, the clone or offspring has the exact same genes as the parent organism.

A clone or a genetic double is not as rare in the natural world as one might suppose. Besides identical twins (who are the result of a fertilized egg separating completely during its two-cell stage), there are numerous examples in the plant kingdom. Almost all potatoes are clones, as are all banana trees grown from root cuttings. For plants, this form of asexual reproduction (an individual copies its genetic material) is known as vegetative reproduction. This is how grass and other plants like strawberries grow and spread. Grass puts out underground shoots, and strawberries send out aboveground runners, both of which eventually form independent, new plants that are genetically identical to the original or parent plant. Most bacteria are also natural clones since they reproduce by a process called binary fission in which they basically split in two, making a pair of identical cells.

Besides these natural types of cloning, a recently developed artificial type of cloning occurs when a segment of deoxyribonucleic acid (DNA) is duplicated outside the body of a plant or animal. Advances with this type of research in which exact copies of DNA segments were made eventually led to scientists being able to clone a complex organism. For example, in 1968, the English biologist John Gurdon cloned a frog by replacing the nucleus of a frog egg cell with the nucleus (a cell's control center) of a cell from another frog's embryo. The egg cell matured into an exact identical twin of the tadpole embryo. Following this success, biologists attempted to clone mice and white rats, but most of the clones did not survive. Cloning mammals proved to be even more difficult and inefficient, with most attempts failing because the cell taken from the embryo was too mature. Its cells had already begun to specialize, as some started making cells for different organs and others making skin cells and limb cells. Overall, it proved very difficult to obtain a mammal embryo cell in its earliest stages of development.

Dolly the sheep was the first mammal to be cloned from the cell of an adult instead of an embryo. Although a scientific breakthrough, the ability to clone an adult mammal has raised many issues. (Reproduced by permission of Archive Photos, Inc.)

This problem was solved on July 5, 1996 when a sheep named Dolly was born in Edinburgh, Scotland. In a dramatic breakthrough, the Scottish embryologist Ian Wilmut was able to clone a mammal from a cell taken not from an embryo but from an adult. His startling success, announced when Dolly was about seven months old, was achieved by Wilmut's unique method of "starving" a cell's nucleus which made it revert

IAN WILMUT

English embryologist (a person specializing in the study of the early development of living things) Ian Wilmut (1944–) produced the first mammal to be cloned from an adult animal. This biological breakthrough meant that future cloned animals might be used to produce large quantities of proteins needed for making certain drugs. It also suggested that these animals might provide a safer organ transplant source for humans.

Ian Wilmut was born in Hampton Lucey, England, the son of a mathematics teacher. He became fascinated with embryology while earning a degree in agricultural science at the University of Nottingham in 1967. Wilmut continued his studies at Darwin College at Cambridge University in England and received a Ph.D. in animal genetic engineering in 1971. He then took a position at the Animal Breeding Research Station in Scotland, now known as the Roslin Institute. While at Darwin College, his dissertation topic was on techniques for freezing boar sperm, and in 1973 he created the first calf ever produced from a frozen embryo. Wilmut continued his research during the 1980s, always with the goal of cloning an animal in mind. A clone is the offspring that results from a form of asexual reproduction. This means that cloning involves only a single parent and does not require the exchange of sex cells from a male and female.

In 1990, Wilmut hired English cell biologist Keith Campbell to work in his cloning laboratory, and it was Campbell's idea that transplanted adult cells had not been working with embryo cells because the two were not "synchronized." Since cells go through specific cycles, regularly growing and dividing and making an entirely new package of chromosomes each time, Campbell argued that adult mammal cells had to be slowed down to be in synch with embryos. Wilmut and Campbell then pioneered a new technique

back to an earlier stage of development. First, Wilmut took unfertilized eggs from an adult female and removed all of its DNA. This left it an empty egg that could still support growth. He then took the udder cells from an adult sheep and raised them in a way designed to "turn off" their specialized genes. One of these donor cells was then fused electrically with the empty egg cell, and the artificially fertilized egg started to divide into an embryo. It was then transplanted into the womb of a sheep, and Dolly, the genetic twin of the animal who donated the udder cell and its own DNA, was eventually born.

The cloning of a mammal produced fear as well as praise among many people, as it raised the possibility of cloning a human being. Biologists tried to ease this fear by pointing out the medical advantages of being

of starving adult cells so they would eventually be in the same cycle as the embryos. Once they "turned off" the specialized adult genes (taken from the udder or milk gland of a six-year-old sheep) and made them act like embryo cells, they fused it with an unfertilized egg that had all of its genetic information-containing deoxyribonucleic acid (DNA) removed. After the artificially fertilized egg started to divide into an embryo, it was transplanted into the womb of a surrogate, or substitute, female sheep where it developed and grew, producing an offspring that was genetically identical to the animal that donated the cell.

Wilmut and Campbell, therefore, produced the cloned lamb named "Dolly" on July 5, 1996. As the first clone from an adult mammal, this successful experiment marked an achievement that some had thought would (or should) never be realized. It also set off a wave of discussion and debate about the implications and ethics of cloning. Naturally, that debate focused on the potential for cloning human beings. While Wilmut remained passionate about his achievement, he stated clearly that cloning a person is ethically unacceptable, and that the primary purpose of his work is to advance the development of drug therapies to combat certain life-threatening diseases. As an example of a health-related product developed from cloning, he offers the possibility of cloning an animal that produces the blood clotting factors that hemophiliacs are lacking. He also envisions organ transplants becoming plentiful and routine by means of inserting a human protein into a cloned animal that allows the animal organs to be more easily accepted by the human patient's body. Wilmut is aware of the ethical concerns many people have about cloning, and he stresses that it is very important to prevent any real misuse if humans are to gain any of cloning potential benefits.

able to clone an animal that contains a certain human gene in its cells. They suggest that such animals could produce a particular enzyme needed by people whose bodies will not produce it, such as the blood-clotting enzyme thrombin, which hemophiliacs lack. However, as with all aspects of genetic engineering, cloning raises many issues with far-reaching social, legal, and ethical implications. These complex issues, in turn, raise many difficult questions, such as who decides what traits are desirable? Are biologists "playing God" by tampering with human DNA? And might a genetic mistake result in some sort of disaster in which a genetic monster like an uncontrollable plague is created?

[*See also* **DNA; Genetic Engineering; Nucleic Acid; Reproduction, Asexual**]

Cnidarian

A cnidarian (ny-DAIR-ee-uhn) is a simple invertebrate (an animal without a backbone) that lives in the water and has a digestive cavity with only one opening. Jellyfish, sea anemones, corals, and hydra are all cnidarians. Cnidarians catch food using armlike, stinging tentacles and their bodies have only two tissue layers, unlike higher animals who have three layers.

While a cnidarian may be more complex than a sponge, it is still a very simple invertebrate. Also called a coelenterate, which means "hollow gut," a cnidarian has been described as having a body that resembles a sack or a hollow bag with a hole in it. This hole is its baglike digestive cavity that has a single mouth or opening at one end. Both food and waste pass through this opening which is usually surrounded by armlike extensions called tentacles. These tentacles are equipped with stinging cells called cnidoblasts that shoot poisonous, spiny threads. Cnidarians use these threads to paralyze and capture small animals that swim into them. The unmoving fish is then pushed by the tentacles through the mouth and into the digestive cavity. Once there, it is broken down and eventually absorbed by cells lining the cavity. There are three main groups of cnidarians: the hydra, the jellyfish, and the corals.

The simplest cnidarian is the hydra, and it is one of the few freshwater cnidarians. Although it is especially tiny, resembling a piece of string on a pond, under a microscope the hydra reveals a hollow trunk with a mouth at one end that is surrounded by a ring of tentacles. It captures tiny animals by stinging them and can stretch when reaching for food. A hydra can reproduce asexually by budding (growing new cells that separate from the parent) or by regeneration (growing a complete organism from a piece).

The jellyfish belongs to a class whose name translates as "cup animals." As an adult, it has an umbrella-shaped or bell-shaped body and swims by pumping water into and out of its digestive cavity. Its mouth is in the center of its underside and has long mouth lobes that hang down from it. The nearly colorless jellyfish we often see while swimming is probably the moon jellyfish. All jellyfish reproduce sexually, joining sperm and egg to produce a larva that first swims around and then attaches itself to the bottom and grows as a polyp. It then breaks off and swims away to assume the shape called a medusa. Since a jellyfish's body is mostly water, when it dies and is washed ashore, it soon dries and leaves only what appears to be a circle of film.

Some other cnidarians do not have the freedom to swim about and spend their entire lives attached to something at the bottom of the ocean, usually with many other of their kind in what are called colonies. These are the coral and sea anemones. Both have tubelike bodies and resemble what some call flower animals. They are brightly colored and gather in clusters that resemble blossoms. Most corals are very small and protect themselves by building hard cases of calcium carbonate (limestone) around themselves. During the day, corals hide inside their shell, but at night they extend their tentacles and catch tiny animals swimming by. When a coral dies, its limestone covering remains and coral reefs are built, formed by countless covering after covering. One large example of a coral reef that has developed over time is The Great Barrier Reef off the coast of Queensland, Australia, which is 95 miles (152.9 kilometers) long.

Community

A community is made up of all of the populations of different species living in a specific environment. A community consists of only the living components (biotic) of the environment. Ecologists study the different roles each species play in their community and also study the different types of communities and how they can change over time. A community has no particular size.

An example of a pond community would be all of the algae, plants, fish, frogs, ducks and other organisms that live in and around a particular pond. These many species living within a community interact with each other in many different ways. These interactions are very important in that they play a significant role in shaping the size and structure of the community. There are three major categories of interactions: competition, predation, and symbiosis.

COMPETITION

Competition occurs when members of the same or different species compete for or share a limited resource. A dramatic example of a competitive interaction affecting a community is the death of or departure of a species from a community because others were using up a scarce resource (like food) that the species needed. Since competition for food almost always exists in nature, each species eventually finds its own "niche" in the community. A niche (also called an ecological niche) is a specific job or role in a community that relates to feeding. In general, the niche of one species does not overlap with that of another species. However, ecologists do not agree as to why this is so. Some say that one species out-competes another in a certain area and that the other species is forced out of the community, thus surrendering the niche to the winner. Others say that one species occupies a certain niche because it is the one best suited (or best physically equipped) to do so. A good example of two species having their own niche are the woodpecker and the nuthatch. Although both these birds eat grubs (insect larvae) found in the cracks of trees, they are seldom in serious competition because each has its own niche. While feeding from the same tree, the woodpecker will begin at the bottom and work its way up. The nuthatch does the opposite, working from the top down.

PREDATION

The second type of interaction between species is predation. Predation occurs when one organism catches and kills another organism. Predation is a good example of how community interactions can result not only in community change but in the actual evolution (gradual genetic change in a group of living things) of the species involved. As a result of one species hunting another, the one doing the hunting may evolve better tools to catch its prey, while the one trying to get away from the predator may evolve better ways of escaping or avoiding being noticed. The numbers of how many predators and prey exist will influence the community structure. If there are many predators in a particular community, the number of prey will probably decrease since many of them are caught and eaten. This could mean that within the prey population there will be less competition among themselves. However, if too many prey disappear, the number of predators is likely to go down, since there will be less and less for predators to eat.

SYMBIOSIS

Symbiosis is the third major interaction that has an effect on a community. Symbiosis describes an especially close relationship between two different species within the same community. Although the term can refer to any type of partnership, it sometimes means a type of relationship that is to the benefit of both species. Flowering plants and bees have a symbiotic relationship that benefits each other. Bees get the nectar they need from flowers and, in turn, the flowers are pollinated by the bees. Certain bacteria have the same relationship with rabbits that cannot digest the cellulose (the walls of a plant cell) in the plants they eat. The bacteria live in the rabbit's gut and benefit from the warm, moist environment; they in turn break down the cellulose for the rabbit to digest.

This interdependence between species can result in evolutionary changes. In the short run, however, a community often experiences a noticeable change in one species when its symbiotic "partner" disappears.

ECOLOGICAL SUCCESSION

Changes that occur over time in a community are called ecological succession. These are usually slow, natural changes that take place in any community. Weather patterns change, as do soil minerals and organisms; even populations rise and fall. Ecologists describe two types of succession, primary and secondary. Primary succession occurs when organisms move into a place that formerly had no life, like a newly formed barrier island. Secondary succession happens after an established community suffers some sort of drastic change, like a fire or volcanic eruption. In this type of succession, there is a pattern by which life introduces itself back into the community. For plant life, a meadow will first develop, to be followed by shrubs and later trees.

Studying communities is more complicated than studying populations since the number and type of interactions can be so large and complex. Even relatively simple communities with small numbers of species form a complicated, interrelated web of dynamic interactions.

[*See also* **Competition; Environment; Population**]

Competition

Competition is a situation that arises when two or more organisms have to share the same limited resources. Competition is a constant in nature since plants and animals almost always have to share such important resources as food, water, space, shelter, sunlight, minerals, and mates. Many ecologists (a person specializing in the study of the relationships between organisms and their environment) consider competition a powerful force that shapes populations (a group of the same species) and communities (groups of different species) as well as the adaptation and evolution of species.

INTRASPECIFIC AND INTERSPECIFIC COMPETITION

There are different types of competition, depending on whether the competing organisms belong to the same species (intraspecific competition) or are from different species (interspecific competition). Intraspecific competition is the most common and often the most fierce since individuals of the same species will have nearly identical needs. The intensity of the competition is also closely related to species population—or how many of the same organisms are in need of the same thing. A rise in population will necessarily result in more competition. In fact, many studies have shown that simple overpopulation in certain species often result in the adults of the species not growing to full size. In other species, overcrowding lowers the number of young produced.

Intraspecific competition can take many forms. One of the more obvious types of competition, called "interference competition," occurs when two competitors of the same species directly confront each other over the same thing. The result can be an actual fight or aggressive poses, displays, and threats. When a hawk aggressively swoops down at another hawk, it is probably engaging in a form of interference competition called territorial competition. Many animals often claim a certain area and will defend it against newcomers entering and trying to take its resources. Males of certain species also engage in a form of interference competition when they fight over who will get to mate with a certain female or with a group of females.

Interspecific competition may not be as fierce as intraspecific competition since no two species ever occupy the same ecological niche (or the precise role that a species plays in its environment). However, the closer the ecological niche is between species, the more fierce the competition. This natural rule, that two species cannot occupy the same niche, is called the competitive exclusion principle. When niches are similar, a phenomenon known as "resource partitioning" occurs. Thus species may eat the same thing, but their feeding habits may be different enough that they do not interfere with each other. For example, a woodpecker and a nuthatch, who both dig for and eat insect grubs that live under tree bark, can both work the same tree without interfering with each other since the woodpecker eats from the bottom up and the nuthatch from the top down.

Competition could also be described as one of the engines driving evolution, since competition is at the core of the concept of natural selection. In the natural world, living things that are not properly "fitted" or suited to their environment are eventually weeded out because they will fail to survive. The saying "survival of he fittest" originated from this phenomenon, since nature allows only those organisms that are best adapted to survive when its resources are limited. Only the best competitors survive and get to reproduce, passing on to their young the characteristics (adaptations) that made them better competitors. Poor competitors seldom survive. Although competition ensures the survival of the fittest, conditions can change, and an individual that survived well in a certain environment may be destroyed if the environment changes, or if the individual moves to a different environment.

[*See also* **Community; Population**]

Crustacean

A crustacean is an invertebrate (an animal without a backbone) with several pairs of jointed legs and two pairs of antennae. It is covered by a tough exoskeleton (a hard outer support structure) with overlapping plates that thin out at the joints to allow maximum movement. Its body is divided into two main regions that are fused together. Most crustaceans live in water.

A crustacean is a member of the phylum Arthropoda, the largest and most successful phylum in the kingdom Animalia. It is also a member of the class Crustacea which is one of the three major groupings of arthropods (the other two are Arachnida and Insecta). The name crustacean is derived from the Latin word *cursta* meaning "crust," and refers to the hard outer shell that this class of invertebrate wears. There are about 40,000 species of crustaceans, including the better-known animals like shrimps, lobsters, crayfishes, and crabs as well as barnacles, water fleas, and isopods like the wood louse. Some are predators and eat other invertebrates, while others are herbivores and eat only plant material. While there are simple crustaceans, most usually have a large diversity (and a large number) of paired appendages (like legs, arms, or pincers). To be classified as a crustacean, an animal must have two joined body parts—a cephalothorax (a head and middle region) and an abdomen (the lower part of the body). The head has two compound eyes that are located on the ends of retractable and flexible stalks. A compound eye is made up of many separate compartments, each having its own lens. A crustacean must also have two pairs of antennae with which it feels and receives chemical stimuli. It must also have at least four pairs of walking legs, and often has more. Shrimp, lobster, crabs, and crayfish are called decapods because they have ten legs. All have a broad, paddle-like tail used for swimming.

As with a representative crustacean species like the crayfish, its first set of legs are adapted as claws. It uses these claws or pincers to obtain food and to defend itself. The other four pairs of smaller legs are used for walking. Behind these walking legs and attached to the lower half of its body called the abdomen are tiny appendages called swimmerets used for swimming and during reproduction. The respiratory system of a crayfish consists of gills over which water passes as the animal moves. The gills are really feathery outgrowths located on both sides of its body. The crayfish has a heart that moves its blood through arteries. Other crustaceans have variations of these systems. For example, a crab has an especially strong claw used for tearing up seaweed and attacking another animal. The barnacle, which attaches itself to a rock by a long stalk, has appendages almost like feathers that are used to comb or sift the water for microscopic food.

The smallest crustacean might be the 7,500 species of copepods, some of which are no more than a few millimeters long. These free-swimming herbivores play an important role in the diet of many fish. All crustaceans reproduce sexually and develop through a series of larval stages.

A crayfish at the Fish Point State Wildlife Area in Michigan displaying its claws. These claws are common traits for crustaceans and are used to catch food and defend themselves. (Reproduced by permission of Field Mark Publications. Photograph by Robert J. Huffman.)

Cytoplasm

The term cytoplasm refers to the contents of a cell, excluding its nucleus. More specifically, cytoplasm refers to the jelly-like or semisolid fluid that is enclosed by the cell's plasma membrane. Before scientists had knowledge of what was contained in a eukaryotic cell (one with a nucleus and an outer membrane), the term cytoplasm was a convenient description for the cell's contents. Now it is known that cytoplasm contains the cell's organelles—the many tiny structures that each have a particular function to perform. The cytoplasm, along with the nucleus, make up what is called the protoplasm, or living material of a cell.

Cytoplasm is the fluid environment in which the cell's metabolism (the chemical processes that make a cell a living thing) takes place. Cytoplasm is a gel-like fluid rich in proteins, fats, carbohydrates, salts, and other chemicals. Unlike a common gelatin, however, cytoplasm is constantly moving and transporting materials from one place to another. This cytoplasmic movement can best be observed in slime molds, amoeba (uh-MEE-buh), and certain species of algae, in which an ordinary light microscope will reveal what appear to be streams of cytoplasm coursing through the interior of a cell.

In addition to these proteins and enzymes in fluid form, the cytoplasm contains a wide variety of organelles. Each of these tiny structures carries out a particular task that is important in maintaining the life of the cell. Some break down food while others move waste around and get it ready to be expelled. Others may store important materials. All organelles in the cytoplasm are surrounded by membranes. Some of the more important organelles found in a cell's cytoplasm are mitochondria (energy generators), ribosomes (assembly units for proteins), endoplasmic reticulum (material transporters), Golgi bodies (storage), and other components like the cell's coded plans and instructions that are carried in its ribonucleic acid (RNA). Plant cells have chloroplasts in their cytoplasm, enabling them to convert sunlight into food. Although animal cells do not have chloroplasts, they do have lysosomes which enable them to digest the food they take in.

[*See also* **Cell**]

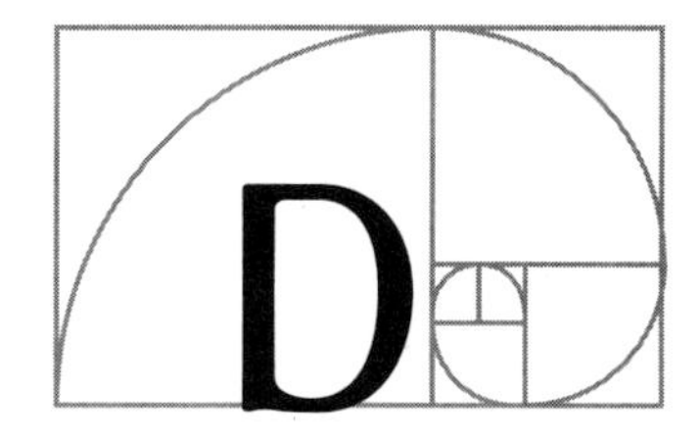

Decomposition

Decomposition is the process by which dead organisms and their wastes are broken down into an organic form that is usable by other organisms. In ecological terms, the chemicals out of which living things are made were only borrowed from the earth, and when they die, they are returned by the process of decomposition. These chemicals are then recycled in order to be used by other living things.

Decomposition could be called rotting or decay, and it is carried out by a very important group of organisms called decomposers. Decomposers are bacteria and fungi as well as other small animals called detritivores. These decomposers break down waste and dead matter into smaller and smaller pieces until all the chemicals they contain are released into the air, water, and soil. Decomposers may be tiny in size but they are huge in numbers. In addition to bacteria and fungi, which are the most important decomposers, such larger animals as the slug, snail, earthworm, woodlouse, and centipede also play an important role in breaking down organic matter. A common example of a fungus and decomposer is a mushroom growing out of a dead tree trunk and feeding on the tree's decaying remains.

Decomposition is an essential stage in the cycling of nutrients through nature's food web (the connected network of producers, consumers, and decomposers). It is nature's way of recycling nutrients so they can be reused. Some of the major nutrients that are recycled include nitrogen, phosphorous, carbon, and oxygen. The decomposers that are a key link in nature's food web because they allow nutrients to be cycled through it continuously. If it were not for decomposers, the food web could not be self-sustaining and would break down because it would eventually have to obtain more nutrients from outside the ecosystem (an area in which living things interact with each other and the environment). For example, without the carbon dioxide that is released by decomposition, green plants would eventually die since they use it to make their food. Without green plants, there would be much less oxygen, since they give it off as part of the process of making their own food. No plants and no oxygen means that all animals would starve. Finally, without decomposition, the world would be a mineral-deficient land full of waste and corpses.

[*See also* **Bacteria; Fungi**]

Desert

A desert is generally a very hot, barren region on Earth that receives little rainfall. Most sources describe a region as being a desert if it receives less than 10 inches (25.4 centimeters) of rain a year. It has also been described as a place where more water evaporates than falls as precipitation. Despite being an extremely harsh environment, deserts support a diverse community of both plant and animal life. As one of the six terrestrial (land) biomes (particular types of large geographic regions), deserts cover between one fifth and one quarter of Earth's surface.

A desert is a stark, dramatic place whose topography (surface conditions) is almost immediately recognizable. Its miles of sand dunes or endless stretches of flat, featureless sand are not easily forgettable; nor are its strangely adapted plants (like cacti) apt to be confused with vegetation from some other region. It is easy to understand what makes a desert what it is. Any part of Earth that constantly experiences a water "debt" rather than a water "surplus" is so dry that the need to capture, conserve, and store water is not only overwhelming, but affects and determines everything living in that place. Despite the impression that a desert is a lifeless place, it is home to certain plants and animals who have adapted to its harsh conditions and who do very well there.

THE LOCATION OF DESERTS

Most of the world's deserts are located on two desert belts that wrap around Earth's equator (the circular band around Earth's middle which divides the Northern and Southern Hemispheres). The belt in the Northern Hemisphere is along the tropic of Cancer and includes the Gobi Desert in China, the Sahara Desert in North Africa, the deserts of southwestern North America, and the Arabian and Iranian deserts in the Middle East. The belt in the Southern Hemisphere is along the tropic of Capricorn and includes the Patagonia Desert in Argentina, the Kalahari Desert of southern Africa, and the Great Victoria and Great Sandy Deserts of Australia. Altogether, there are about twelve major deserts, the largest of which is the Sahara Desert which measures 3.5 million square miles (5.63 million square kilometers). This is an area as big as the entire United States.

THE CREATION OF DESERTS

In a way, deserts are made and not born, meaning that Earth's weather patterns created a desert in the first place and continue to work to keep it that way. These regular patterns, or moving currents of hot and cold air interact with each other so that descending currents of air pick up moisture and dry out the land. Mountain ranges also influence these currents, as dry air moving off their slopes evaporate even more moisture. The steady lack of moisture in the air above a desert region leads to extreme changes in temperature once the Sun goes down. In normally humid areas, the moisture in the air acts as an insulating barrier, and clouds keep some of the daytime warmth from the Sun trapped, thereby moderating temperatures. However in a desert, which has no moisture in the air above it, there are no clouds to act as a blanket, meaning that although daytime temperatures are extremely hot, they can be near freezing at night.

As with any biome, deserts vary considerably throughout the world, and they can be as diverse as the lifeless-looking and appropriately named Death Valley in California and Nevada, and the almost-lush looking Vazcaino Desert in Mexico when it bursts into flower following its annual spring rain. Even in as harsh an environment as Death Valley or the Sahara Desert, life can be found. Sometimes life is a dormant seed buried for years and waiting for a bit of moisture so that the seed can spring into existence as an aboveground plant. Other times desert life is a toad hibernating below ground and rushing to find a mate and lay its eggs as soon as it rains. Life in a desert is a constant challenge, and plant and animal inhabitants do not have the luxury of being wasteful that other organisms in more temperate climates might have.

PLANT LIFE IN THE DESERT

Desert plants have evolved many methods to obtain and efficiently use available water. Certain ones compress their entire life cycle into one growing season. The seeds or bulbs of some flowering desert plants can lie dormant in the soil for years until a heavy rain enables them to germinate (sprout), grow, and bloom. Woody plants may develop deep root systems—like the mesquite whose tap root can measure 50 feet (15.24

meters) down although the aboveground tree is only about 10 feet (3.048 meters) tall. They may also develop a network of shallow, spreading, hair-like roots that can take up water from dew and the occasional rain shower before it seeps below ground. For many plants, the answer to years of absolute drought is to drop leaves, allow the aboveground part of the plant to die, and keep the underground root alive in a state of dormancy (functioning slowly or not at all). Conserving and storing water becomes important for a plant once it has obtained moisture. Since all plants lose water by evaporation from their leaves, many desert species minimize this by having very small or rolled leaves, or by turning their leaves into spines or barbs. These thornlike leaves protect a plant's water supply from animals. The problem of storing water is solved by the cactus, which is a succulent and can store water in its leaves, stems, and roots. An amazing example of adaptation is the Saguro cactus of the American southwest. The trunk of this large cactus is folded or pleated like an accordion, which can unfold and expand as the plant absorbs water after a heavy rain. A Saguro that is 20 feet tall (6.1 meters) can hold more than one ton (1.102 metric tons) of water.

Cacti in the Sonoran Desert in Arizona. The cactus is specially adapted to life in the dry desert. (Reproduced by permission of JLM Visuals.)

ANIMAL LIFE IN THE DESERT

Desert animals, like desert plants, have also evolved ways to cope with the desert's arid environment using avoidance and/or adaptation. Besides the highly adapted camel, most desert animals are small and do not have an extensive range. While their size limits their ability to leave, it does make easier their ability to remain in cool underground burrows during the day and emerge only after dark to feed. Animals that do this are called "nocturnal" or "crepuscular." Other small mammals and reptiles survive the most extreme times by a process called estivation, which is similar to the sleeplike state of hibernation. Other animals have adapted specialized body parts to help them cool off. A well-known example is the huge ears of the small Fennec fox of the Sahara Desert and the Kit fox of North America. Both have enormous ears whose dense network of tiny blood vessels run just below their skin and act as radiators, releasing excess heat. Larger desert animals developed broad hooves that allow them to move more easily over soft sand. Some animals can actually slow down their production of body heat by varying their heart rate, while others reabsorb the water in their urine several times before finally excreting a highly concentrated form of urine.

Just as the animal and plant population in deserts is not overly abundant because the desert's difficult conditions can only support small numbers, deserts cannot support humans in large numbers. People must, like animals and plants, make adjustments in order to survive in the desert's extremes, and in the past they have lived in mud houses that kept cool in the daytime and provided warmth at night. Long robes were often used in Africa and the Middle East for protection against the scorching sunlight and blowing sand. With today's technology, however, people can live comfortably in a desert if they have air conditioning and an adequate water supply. A good, steady source of water also allows humans to raise crops in a desert since they are usually naturally fertile regions because there is seldom enough water to leech away important nutrients. Crops can be grown on desert lands with irrigation, but farmers must be prepared to deal with a buildup of salts in the soil as a result of evaporation (which takes away most of the water they put down). Humans can also be responsible for creating deserts or allowing an existing desert to spread. This is usually the result of burning or overgrazing of animals. When a desert encroaches, or spreads, to arable land (land able to be farmed), that process is called "desertification."

[*See also* **Biome**]

Diffusion

Diffusion is the movement or spreading out of a substance from an area of high concentration to the area of lowest concentration. Diffusion takes place at the cellular level in both living and nonliving things. Simple animals that do not have an internal circulatory system rely on diffusion to exchange gases and obtain nutrients. The root cells of green plants obtain their water from the soil by diffusion. Diffusion requires no output of energy on the part of the cell.

Diffusion is a natural phenomenon that happens under certain conditions and occurs at the molecular level. Because molecules are constantly moving, their natural tendency is for different types to mix with one another. This movement of molecules is random and depends on the amount of energy (called kinetic energy) in each molecule. As different types of molecules move about and mix together, the only pattern noticeable is their overall movement from an area of high concentration (where they are all together) to an area of lowest concentration (where there are the fewest of them). While an individual molecule may not do this exactly, the net or overall movement of the group of molecules will always move in that manner. This net movement is called diffusion.

A good example of everyday diffusion is the way tobacco smoke spreads throughout the still air of a room. Perfume does the same thing in the motionless air of a room. For the life sciences, diffusion takes place at the cellular level of both plants and animals. Since cell membranes are composed of molecules that are always in motion, there are always temporary openings in cell membranes. Living cells are always bathed in liquid. If the liquid concentration of a certain type of molecule is higher on the outside of a cell membrane than on the inside, those molecules will

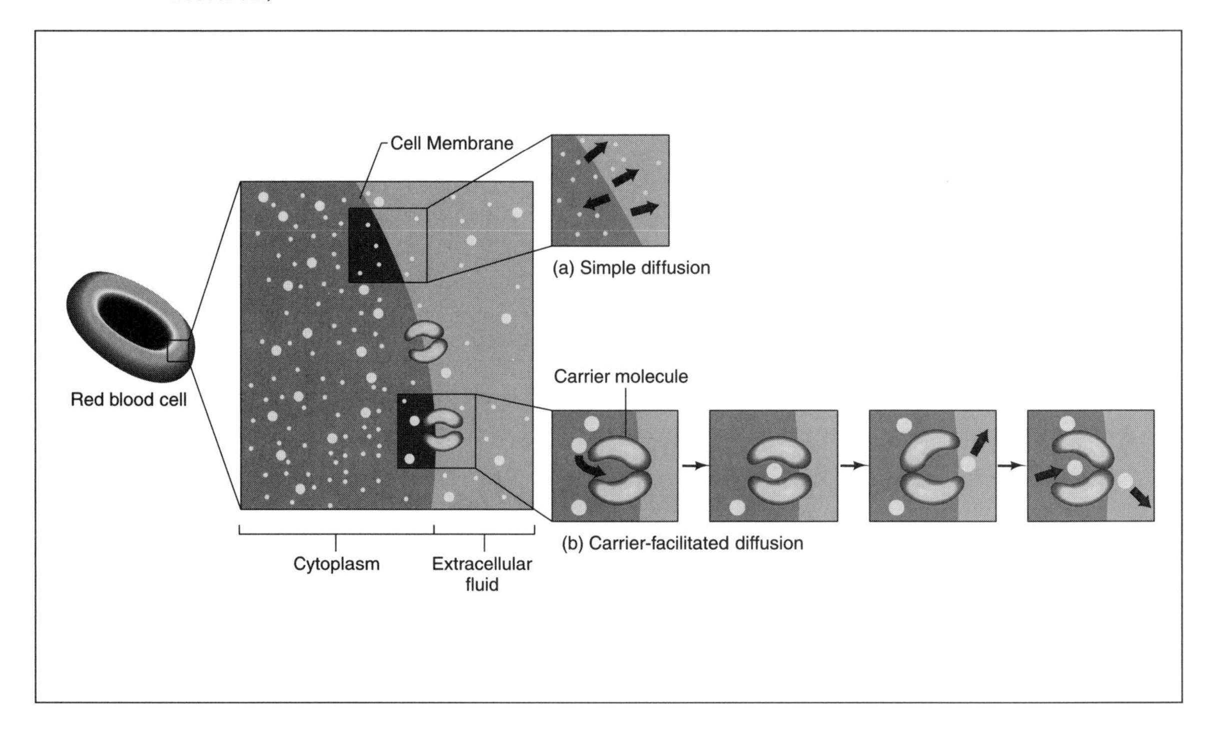

An illustration of diffusion in a red blood cell. (Illustration by Hans & Cassidy. Courtesy of Gale Research.)

diffuse through the membrane into the cell. The opposite can also occur if the liquid concentration is higher on the inside of a cell. The rate of diffusion depends on the unevenness of the distribution and the size of the molecules. When the molecules on either side of the membrane are equally distributed, no diffusion will take place.

The process of diffusion is an important method for cells to be able to exchange substances with their environment. As a result, the cells are able to refill what they have run out of and dispose what they have too much of. However, since the cell has no real control of this process, it is called passive diffusion. Diffusion also takes place in the lungs of animals. When blood enters the lungs to fill up with oxygen, it is carrying carbon dioxide, a waste product, in high concentration. Immediately, the carbon dioxide diffuses from the blood into the lungs and oxygen diffuses from the lungs into the blood.

This same exchange takes place in single-celled animals and in many simple aquatic animals as well. Animals such as sponges, flatworms, and hydras rely on diffusion to get oxygen and food from the surrounding water and to remove waste. Such simple animals can do this either because their bodies have many openings (like sponges) or because they have body walls that are only two or three cells thick (like hydras). Diffusion can work in such animals since they have simple body parts and minimal demands. However, creatures with fast metabolisms (all the chemical processes going on in a living thing) need real circulatory systems that can work fast and transport large amounts. Plants with vascular systems (internal pipelines for transporting food and water) also use diffusion to take water and oxygen into their roots. For example, if the cells of a root have only a little oxygen in them and the surrounding soil has a great deal, oxygen molecules will automatically move from the soil to the root. As with animal cells, the plant does not expend any energy to accomplish this.

[*See also* **Respiratory System**]

Digestive System

A digestive system is a system that allows an organism to take in food, break it down, absorb its nutrients, and excrete what is not usable. All organisms that cannot internally make their own food (as plants do) must ingest or eat it, and therefore must have a digestive system in some form. Different types of animals have different digestive systems according to their main diet and the amount they eat.

All living things need food in order to continue to live, grow, and reproduce. Except for green plants and some algae that can make their own food using the Sun's energy, all other living things get their energy from eating other living things, such as plants or animals. However, these plants and animals used as food are made up of large molecules that cannot be used by an organism's cells unless they are changed into smaller molecules that can be absorbed. The entire process by which food is converted into a form the body can use is called digestion. This process of digestion is carried out by the organism's digestive system. Digestive systems range from the very simple, primitive systems of one-celled organisms to the complex, many-organ systems used by vertebrates (animals with backbones).

INTRACELLULAR DIGESTION

Very simple, single-celled organisms practice what is called intracellular digestion in which they engulf or surround a food particle with their outer membrane. During this type of digestion, these organisms literally bring the food particle inside the cell. Strong enzymes (proteins that control the rate of chemical changes) break the food particle down into its usable components, which are then absorbed into the cell's cytoplasm (the jelly-like fluid inside a cell). Waste products are packaged up and passed back out through the cell membrane.

EXTRACELLULAR DIGESTION

Other slightly more complex organisms like a sponge may have a mouth that leads to a large, open body cavity. Organisms that have only one opening through which passes both their food and their waste are said to have an incomplete digestive system. Flatworms and hydras have this type of digestive system, sometimes called a blind gut. Food enters its mouth and is partially digested by chemicals released into its gut. This is called extracellular digestion because it occurs inside the gut cavity and not inside a cell. Once the food has been broken down, the smaller bits can be absorbed by the cells that line the gut. Waste products are passed back out through the mouth. Because of this two-way traffic, the organism's cavity cannot be subdivided into specialized compartments.

Opposite: A labeled diagram showing all of the parts of the human digestive system. (Illustration by Hans & Cassidy. Courtesy of Gale Research.)

More complex organisms have more complicated digestive systems. These are called complete digestive systems. Higher up the evolutionary ladder, the blind gut develops a separate opening for waste removal, called the anus. This is seen in earthworms, clams, crabs, spiders, and starfish, among others, who have the simplest form of a complete digestive system. Food enters the mouth, is broken down, and passes in one

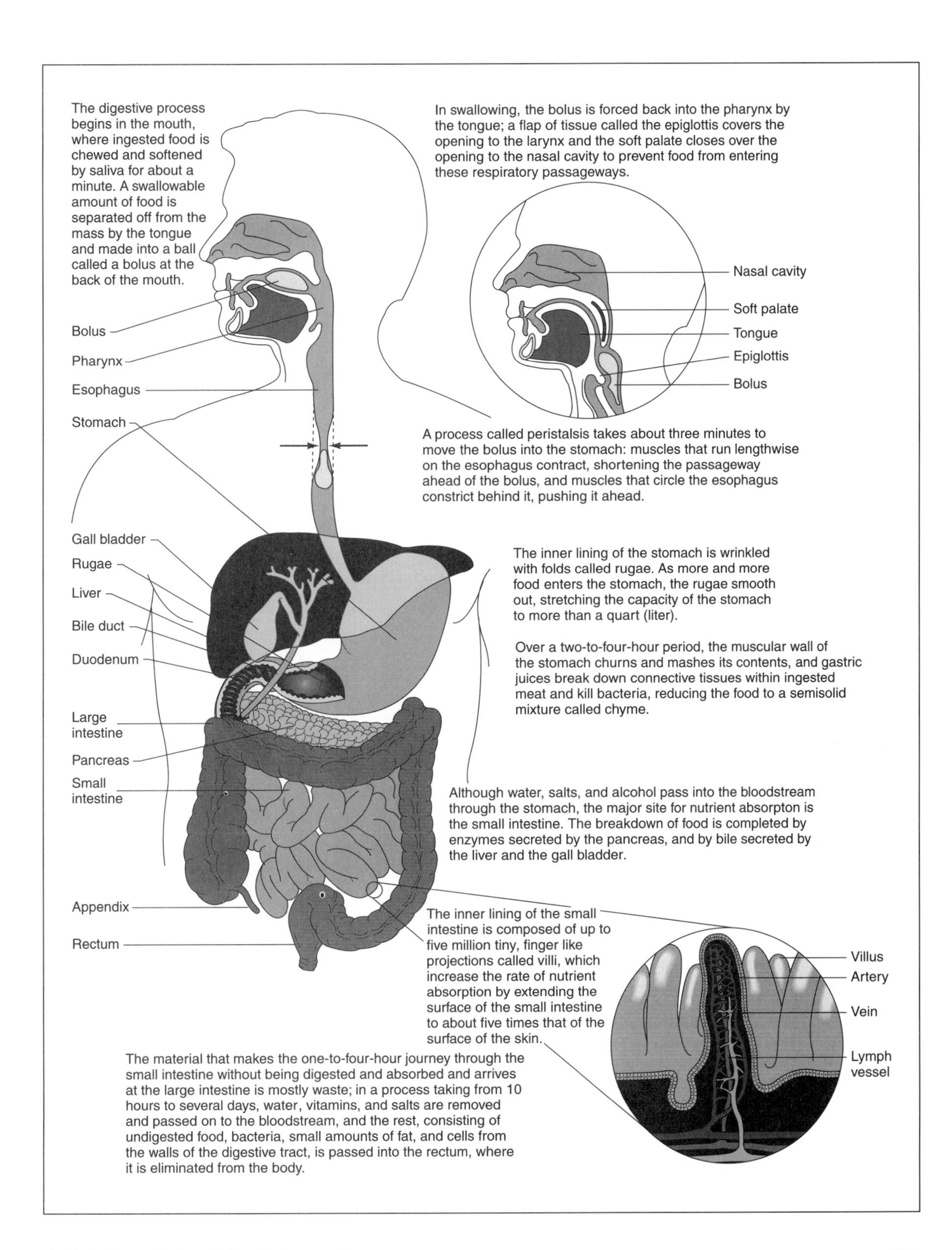

The digestive process begins in the mouth, where ingested food is chewed and softened by saliva for about a minute. A swallowable amount of food is separated off from the mass by the tongue and made into a ball called a bolus at the back of the mouth.
In swallowing, the bolus is forced back into the pharynx by the tongue; a flap of tissue called the epiglottis covers the opening to the larynx and the soft palate closes over the opening to the nasal cavity to prevent food from entering these respiratory passageways.
Bolus
Pharynx
Esophagus
Stomach
Nasal cavity
Soft palate
Tongue
Epiglottis
Bolus
A process called peristalsis takes about three minutes to move the bolus into the stomach: muscles that run lengthwise on the esophagus contract, shortening the passageway ahead of the bolus, and muscles that circle the esophagus constrict behind it, pushing it ahead.
Gall bladder
Rugae
Liver
Bile duct
Duodenum
The inner lining of the stomach is wrinkled with folds called rugae. As more and more food enters the stomach, the rugae smooth out, stretching the capacity of the stomach to more than a quart (liter).
Over a two-to-four-hour period, the muscular wall of the stomach churns and mashes its contents, and gastric juices break down connective tissues within ingested meat and kill bacteria, reducing the food to a semisolid mixture called chyme.
Large intestine
Pancreas
Small intestine
Although water, salts, and alcohol pass into the bloodstream through the stomach, the major site for nutrient absorpton is the small intestine. The breakdown of food is completed by enzymes secreted by the pancreas, and by bile secreted by the liver and the gall bladder.
Appendix
Rectum
The inner lining of the small intestine is composed of up to five million tiny, finger like projections called villi, which increase the rate of nutrient absorption by extending the surface of the small intestine to about five times that of the surface of the skin.
Villus
Artery
Vein
Lymph vessel
The material that makes the one-to-four-hour journey through the small intestine without being digested and absorbed and arrives at the large intestine is mostly waste; in a process taking from 10 hours to several days, water, vitamins, and salts are removed and passed on to the bloodstream, and the rest, consisting of undigested food, bacteria, small amounts of fat, and cells from the walls of the digestive tract, is passed into the rectum, where it is eliminated from the body.

WILLIAM BEAUMONT

American surgeon William Beaumont (1785-1853) conducted pioneering studies on how the human stomach works. His highly accurate and well-documented firsthand observations of the stomach proved that digestion is basically a chemical process.

If U.S. Army surgeon William Beaumont had not been in the right place at the right time, he would never have become a pioneer of science. Until the event occurred that changed his life, Beaumont had done nothing remarkable. Born in Lebanon, Connecticut, on a farm, he was able to study medicine at St. Albans in Vermont. By 1812, he became an assistant surgeon in the U.S. Army. Three years later, he left the army and tried private practice in Plattsburgh, New York. However, in 1820 he returned to the army and was eventually transferred to the frontier post of Fort Mackinac in Michigan. While there, a bizarre medical event occurred on June 6, 1822. That day, Beaumont was called upon to assist a patient who had been accidentally shot at close range with a shotgun. What Beaumont found was a young French-Canadian trapper named Alexis St. Martin who had been struck by the blast on his left side. The shotgun had taken a deep chunk out of his side and no one expected the young man to live through the night. Beaumont tended him with care and skill nonetheless, and to everyone's surprise he remained alive. Beaumont continued his care, changing his bandages every day for a year, and in time, St. Martin was fully recovered.

However, despite the fact that he had regained his full strength and seemed normal, St. Martin was not literally whole. His wound never fully closed, and he was left with an inch-wide opening through which Beaumont could insert his finger all the way into St. Martin's stomach. The trapper had been left with what is called a permanent traumatic fistula, meaning he had a hole in his side that led directly to his stomach. About a year later when

direction through a straight digestive tube where it is absorbed. Waste passes out of the body through an anus at the organism's other end. Earthworm waste called casts are deposited on the soil's surface, adding needed nutrients.

HOW THE DIGESTIVE SYSTEM WORKS

As organisms became more complex and evolved into what are called the higher animals, their digestive systems developed an alimentary tract or alimentary canal with specialized structures and compartments. Among vertebrates, a basic digestive scheme came about that resulted in structures responsible for: receiving food, conducting and storing food, break-

St. Martin needed to take some medicine from Beaumont, the surgeon decided to try an experiment and administered the medicine directly into the stomach rather than orally as he would have done normally. Beaumont soon found that the medicine worked on St. Martin exactly as it would have if it had been administered the regular way. This led Beaumont to realize that he had a unique research opportunity at hand, and he soon began a series of experiments and observations on his subject. First Beaumont attached small chunks of food to a string and inserted them directly into the St. Martin's stomach. Beaumont would then withdraw the string and observe the results of digestion hour by hour. Later, by using a hand lens, he began actually looking into his patient's stomach to see how it behaved. He was also able to extract and analyze samples of gastric juice and stomach contents, establishing that digestion is indeed a chemical process. He could also observe the muscular movements of the stomach. Over the next few years, Beaumont conducted more than two hundred carefully detailed experiments and, in 1833, published his findings in *Experiments and Observations on the Gastric Juice and the Physiology of Digestion.* As the first, well-documented and accurate observation of the digestive processes of a living human being, Beaumont's book was a one-of-kind source on the process of digestion. It also suggested to some scientists the possibility of using artificial fistulas on their research animals as a way of learning more about their bodies.

The human side of this story should also be noted. By about 1835, St. Martin had had enough, not only of Beaumont's experiments, but also of the bullying Beaumont himself. Poor St. Martin refused to cooperate anymore and eventually just ran away from Beaumont for good and returned to his native Canada. In his own way, the wounded trapper who lived to the ripe old age of eighty-two, had made his own contributions toward understanding more about the human body.

ing food down and absorbing its useful nutrients, and absorbing water while eliminating wastes. The alimentary canal in vertebrates is able to move food from the mouth to the anus entirely on its own. It does this by contracting its circular muscles and pushing food further along. This happens in a rhythmic wave called peristalsis. The canal itself is constantly lubricated by mucus, making the food move easily when the circular muscles contract. Automatically, these important muscles are regularly contracting and relaxing, pushing food and waste along as if a fist were closing on a tube, squeezing its contents in a certain direction.

In all vertebrates, food encounters four major areas in the alimentary canal where it is received and ingested (swallowed), broken down, absorbed, and eliminated. The mouth is responsible for receiving food, and it is usually designed with the animal's diet in mind. The mouth may have teeth which, helped by the tongue, mechanically break food down into smaller pieces. Some species, like snakes, smell with their tongue, while others, such as frogs, capture their food with it. Since mammals chew their food in their mouth, they also have salivary glands that secrete enzymes that moisten their food and begin the chemical breakdown process. The esophagus connects to the stomach and moves the food downward after it is swallowed. Unlike most vertebrates, however, birds have an enlarged esophagus called a crop that is used to store food before it is sent to the stomach. It also can produce a "milky" substance that a parent bird can regurgitate (spit back up) and feed its young.

In most vertebrates, the real grinding and digestion of food is carried out in the stomach. This is a muscular pouch that works to mix the food with a highly acidic combination of chemicals that break it down or dissolve it further. Powerful stomach muscles churn up the food while certain cells lining the stomach produce gastric juice (mainly hydrochloric acid) that turn the broken-down food into a milky substance known as chyme. The stomach itself is protected from the strong acid by a thick layer of mucus. As the stomach fills with chyme, it gradually releases small amounts into the duodenum, the top part of the small intestine. The stomach of some vertebrates, like birds, who swallow their food without chewing (like birds), are called gizzards and often contain pebbles that mechanically break up the food for them as their muscle walls contract. Some plant-eating mammals called ruminants, like cows and horses, have special stomach chambers that help break down their difficult-to-digest diet. They also must have certain microorganisms in their gut in order to break down their food since mammals do not produce the digestive enzymes needed to break down cellulose (tough plant walls).

Once the now-liquid food passes into the duodenum, it begins its final trip to becoming completely digested and prepared for absorption. At this point, the liver and the pancreas come into play as both have ducts leading into the duodenum. The pancreas secretes enzymes or pancreatic juice that is alkaline and counteracts the strong acid made by the stomach. It also breaks down food molecules in the duodenum. The liver produces a fluid called bile that is essential if the body is to digest fats. Bile emulsifies (breaks down) large fat globules so they can be absorbed. Without bile, which is stored in a sac called the gall bladder, most of the fat would pass through the digestive system undigested. From the duodenum, the digested food flows into the small intestine, called the ileum in humans. The small intestine has densely folded structures called villi that look like tiny tongues or fingers. These villi absorb the molecules of nutrients and then rapidly send them into the bloodstream. The parts of food that have not been digested or absorbed continue to pass down into the large intestine (called the colon in humans) in the chyme. The large intestine's job is to absorb nearly all its water, and what is left arrives at the rectum where it is stored as feces. When a sufficient quantity has accumulated, the feces are expelled through the anus. This is called elimination.

Although mammals all have the same basic digestive system, there are differences that reflect their diet. Carnivores, or meat-eaters who have special teeth and live by hunting, have fairly short digestive systems since meat is mostly protein and requires very little hard work to digest. Some carnivores swallow their prey whole and thus do not need to have salivary glands. Herbivores (plant eaters) have to consume huge amounts of vegetative matter that is also hard to break down. As a consequence, their digestive systems are much more complicated than that of a carnivore. For example, a cow has four stomachs and an extremely long and coiled intestine. Some animals, like rodents and rabbits, even reingest their waste pellets so that their food passes through their systems twice. Since their digestion is basically incomplete, they must take in their partially digested fiber and have it pass through their digestive system again to benefit from it fully.

Dinosaur

A dinosaur is an extinct vertebrate (an animal with a backbone) reptile. The first dinosaurs appeared on Earth around 220,000,000 years ago and after surviving for 140,000,000 years, suddenly disappeared. Certain dinosaur species, like the *Brachiosaurus,* were the largest animals ever to have lived. All of the knowledge about dinosaurs is the result of studying the fossil remains that have been discovered in all parts of the world.

Dinosaurs lived during a time in Earth's history called the Mesozoic Era, also called the Age of Reptiles. In many ways, dinosaurs were much like the reptiles we know today—the familiar snakes, turtles, lizards, and crocodiles. Like them, dinosaurs may have been ectothermic or cold-blooded, meaning that their internal body temperature would change according to the temperature of their environment. However, the great physical bulk of some species suggests that it would have taken them a very long time to reach their full size, since ectothermic animals grow very slowly. Like today's reptiles, they varied greatly in size, from those the size of a chicken to others that grew to more than 90 feet (27.4 meters) long. Some dinosaurs were carnivores (meat-eaters) and others were herbivores (plant-eaters). As a reptile, they laid eggs that had tough outer shells and that may have contained enough water and food for the dinosaur embryo to grow and finally hatch.

KNOWLEDGE OF DINOSAURS REVEALED THROUGH FOSSILS

Although a great deal more is known about dinosaurs today compared to when the first fossil bones were discovered in England around 1822, new facts are learned every year as new finds reveal more about their lives and habits. One thing that is known is how dinosaurs actually reproduced. It is known, however, that many probably laid eggs as all reptiles do. Also, it is not known how long they normally lived. Some species may have lived in herds, while others could have been solitary. Another mystery is whether plant-eaters ate underwater plants or leaves on trees, or if carnivores ate other dinosaurs. Among the many things not known

A duck-billed dinosaur fossil that is 78,000,000 years ago. Fossils have been a tremendous help to scientists in learning what dinosaurs were like before they became extinct. (Reproduced by permission of Photo Researchers, Inc.)

about dinosaurs, certainly the biggest and most important of all is the actual reason for their sudden and total disappearance. It is known that after entirely dominating the Earth, dinosaurs went extinct about 63,000,000 years ago.

The fossilized remains of dinosaurs have been found in all parts of the world because at the time they lived, there was a single land mass or continent called Pangaea. When this single continent slowly broke up and moved apart into the several ones evident today, the fossil remains were scattered along with the land masses. This fossil evidence is available today only because of a particular set of circumstances that occurred millions of years ago. If a dinosaur was stuck in soft mud and died there, or if it died and simply fell in and sank, it would sometimes be covered by more and more sediment (that was moved there by wind, water, or ice). Over even more time, these sediments would be compressed by layers of Earth deposited on top of it until everything was transformed by great pressure into solid rock. After nearly 200 years of collecting and studying fossilized dinosaur bones, fossilized eggs, and footprints left in rock, scientists have been able to reconstruct several species with some degree of certainty. They have also been able to classify them as they would any living animal, and have divided dinosaurs into two main groups according to the structure of their hips. This may sound strange, but when people realize that how their hips and those of animals are shaped and function affects how they move about, it starts to make sense.

DINOSAUR GROUPS

The first group or order called Saurischia had a hip structure that resembled a lizard. The second order, called Ornithischia, had hips that were built like those of a bird. The first group included the largest (and most ferocious) dinosaurs. Thus the well-known Apatosaurus (formerly known as the Brontosaurus) is among this group, as is the fearsome Tyrannosaurus rex. Many believe that this large group consisted mainly of carnivores (meat-eaters), and it is known that they walked mainly on four feet, lived mostly on land, and had barrel-like bodies and legs that looked like columns, as well as long, heavy tails. The second group with bird-like hips are believed to be mainly plant-eaters. The well-known Stegosaurus was a member of this order. It had the smallest brain compared to body size of any dinosaur.

Although there are many things left to learn about dinosaurs to understand their life cycle, habits, and internal functions, what remains the largest gap in the knowledge about these great beasts is the cause of their sudden extinction. Scientists now believe that a mass extinction must have somehow occurred, but they are still not in agreement as to its cause. Scientists have already eliminated theories that say that dinosaurs simply grew too large to hold themselves up. Instead, many now think that dinosaurs were already on the decline when something very big and destructive happened. One theory says that the key event was Earth being struck by a massive piece of debris from space, like an asteroid. This would have caused tons upon tons of dust and soot to clog Earth's atmosphere, causing either prolonged darkness that cooled the planet or a greenhouse effect that trapped warmth and caused the surface to overheat. Others say that climates may have simply changed too fast for dinosaurs to adapt. Whatever the exact nature of the cause or causes, something did happen with which the dinosaurs were unable to cope, and they all eventually disappeared. Science may never know for sure what killed all the dinosaurs.

[*See also* **Evolution; Fossil; Geologic Record; Paleontology**]

DNA (Deoxyribonucleic Acid)

Deoxyribonucleic acid, or DNA, is the genetic material that carries the code for all living things. This code determines the form, development, and behavior patterns of an organism and is part of the chromosomes that exist within the nucleus of cells. DNA consists of two long chains joined together by chemicals called bases and coiled together into a twisted-ladder shape.

DNA is a large molecule found in almost all organisms and contains codes for the making and using of proteins. Since proteins carry out the work of all cells, it is DNA that ultimately controls and directs all the activities of a living cell. Biologists have known about DNA for a very long time. Even before they discovered that genes control heredity, they were aware that the cell's chromosomes were made up of protein and a special chemical they called deoxyribonucleic acid (DNA). Although DNA was discovered in 1869, more than fifty years passed before biologists believed that genes were composed of DNA. As a nucleic acid, DNA was considered too simple a chemical to contain the huge amount of complex information needed to determine heredity. Since it is made of only four or sometimes five chemical bases, called nucleotides, no one thought that DNA was complex enough. However, as more experimental evidence began to point toward DNA as key in the transmission of hereditary characteristics, more scientists began to turn their attention to DNA.

WATSON AND CRICK DISCOVER THE STRUCTURE OF DNA

By the early 1950s, an unusual pair of scientists teamed their efforts in the passionate belief that the structure of DNA held the key to understanding how genetic information is stored and transmitted. In 1951, the twenty-four-year-old American geneticist James Watson met the thirty-six-year-old English physicist (and self-trained chemist) Francis Crick and the two decided to try to solve the puzzle of how such a simple material as DNA could store so much complicated information. Both knew they did not have to make any new discoveries, but instead, had to solve what might be called the "molecular architecture" of DNA.

The key to solving that problem lay in a technique known as x-ray crystallography. When x rays are directed at a crystal of some material, such as DNA, they are reflected and refracted by the atoms that make up the crystal. A refraction pattern is produced from which a skilled observer can determine how the atoms of the crystal are arranged. However, this is more difficult than it sounds, and Watson and Crick were trying to solve a very difficult puzzle. It was the chemical composition of DNA itself that led them to the correct model. They knew that DNA was composed of four chemicals—adenine (A), guanine (G), cytosine (C), and thymine (T)—and that A was always paired with T, and C always with G. Knowing this and studying the x rays led them to realize that this alignment could only happen if DNA was made up of two strands that were twisted together to form what is called a "double helix." (A double helix is the correct name for a corkscrew-like spiral shape.) In March 1953 Watson and Crick built a wire model showing that the DNA molecule could be thought of as a ladder where the nucleotide bases (A, T, G, C) form the "rungs" connecting the two side rails. The sides were then twisted to make the double helix. In Watson and Crick's Nobel Prize-winning work, the base pairs were the critical part that allowed them to explain how nature stores and uses a genetic code. Each DNA base is like a letter of the alphabet, and a sequence of nucleotide bases can be thought of as forming a message. Put another way, each "rung" (base pair) of the twisted ladder consists of some combination of the four chemicals (A, T, G, C) that form the coded message.

THE CELL'S INSTRUCTION MANUAL

DNA has been called the instruction manual for the cell. It has also been called the chemical language in which "gene recipes" are written. This is because genes can be considered recipes for making proteins, and proteins control the characteristics of all organisms. These codes or recipes are written with the four nucleotide building blocks (the bases A, T, G, and C). Each gene has several thousand bases joined together in a precise

JAMES DEWEY WATSON AND FRANCIS HARRY COMPTON CRICK

The team of Watson and Crick discovered the structure of deoxyribonucleic acid (DNA), one of the most important discoveries of modern science. Their model explained how genetic information is coded and how DNA makes copies of itself. This discovery formed the basis for all the genetic developments that have followed.

English molecular biologist Francis Crick (1916–) was born in Northampton, England. As a boy, he was very interested in chemistry, although he eventually obtained a degree in physics from University College in London. During World War II (1939–45) he worked on the development of radar (an instrument used to determine an object's position, speed, or other characteristics) and new weapon design. After the war, Crick decided he wanted to concentrate on biology and did so on his own, eventually taking a job at the Cavendish Laboratory where he would meet the younger Watson.

American molecular biologist James Watson (1928–) was a former "Quiz Kid," which was the name of a popular radio show in the 1940s. Born in Chicago, Illinois, Watson was a child prodigy (an exceptionally smart person) who graduated from the University of Chicago when he was nineteen and who obtained his Ph.D. from the University of Indiana at twenty-two. After receiving a fellowship to study in Copenhagen, Denmark, he joined the Cavendish Laboratory at Cambridge, England, where he met Crick.

Although Crick was twelve years older than Watson, both shared what was described as "youthful arrogance" and found that they got along extremely well both personally and professionally. Crick had been studying protein structure, and Watson came to Cambridge highly interested in discovering the basic substance of genes. With these complementary goals, they teamed up to try to unravel the structure of DNA, the carrier of genetic information at the molecular level.

code that is different from the code for any other gene. For example, a sequence such as A-T-T-C-G-C-T... etc. might tell a cell to make one type of protein (for red hair), while another sequence such as G-C-T-C-T-C-G ... etc. might code for a different type of protein (for blonde hair). When cells reproduce by division (a process called mitosis), each parent cell must make sure that its daughter cell (its exact duplicate) gets a complete copy of its DNA. This is accomplished by a process called "replication" in which the two strands or rails of the DNA ladder "unzip" themselves down the middle of the bases (or rungs of the ladder). Since A always links to T and G always to C, each separate rail of the ladder becomes a

Beginning in 1951, they worked to create a DNA model that would explain how it could copy and pass on its instructions to every new cell in a living thing. After a great deal of experimentation, they recognized the importance of DNA x rays being done by the English biochemist Rosalind Franklin (1920–1958), and soon built an accurate spiral-shaped model, called a double helix, in which they said two parallel chains of alternate sugar and phosphate groups were linked by pairs of organic bases. Their model looked like a twisted, spiral staircase. They then theorized that replication (the process by which DNA molecules copy themselves) occurs by a parting, or unwinding, of the two strands, or bases, of the staircase that then unite with newly created strands to form new DNA molecules (made up of one old strand and one new strand). Watson and Crick then published their findings in the journal *Nature,* which appeared on April 25, 1953 (with Watson's name appearing first due to a coin toss). Other researchers soon confirmed their hypothesis and the Watson-Crick model was accepted as correct.

Watson went on to write several highly regarded books, one of which became the first widely used textbook on molecular biology. He taught at Harvard University, became director of Cold Spring Harbor Laboratory of Quantitative Biology in Massachusetts, and served as the first director of the United States Human Genome Project. Crick joined the Salk Institute for Biological Studies in San Diego, California, in 1977. In 1962, Watson and Crick shared the Nobel Prize in Physiology and Medicine with their colleague, English physicist, Maurice H. F. Wilkins, for their discovery of the structure of DNA. Rosalind Franklin worked on Wilkins's team, and she would have received the award had she lived. The Watson-Crick discovery of DNA structure ushered in the modern era of molecular biology and made possible all that has happened since. There would be no understanding of human genetics without their discovery.

template or model for free-floating bases to link up to. The result is the existence of two identical double helixes where there was just one.

The other important processes to understand are called "transcription" and "translation," which are the two stages of making a protein. In transcription, DNA "unzips" again and another type of nucleic acid called ribonucleic acid (or messenger RNA) uses one strand of DNA as a template to make an exact, single-strand copy. The RNA then leaves the nucleus with its message and becomes a template for the production of protein (by a ribosome in the cell) in what is known as translation. This process is going on all the time in our bodies, since cells are constantly called upon to make protein molecules.

Breakthroughs in understanding DNA have also led to the use of DNA in forensic science where "DNA fingerprinting" or "DNA profiling" is conducted. This use of DNA is based on the fact that repetitive sequences of DNA vary greatly among individuals, since each person has his or her own unique code. It has also led to the beginnings of treatment for hereditary disease by genetic engineering. In theory, doctors will be able to replace a defective (inherited) gene that causes a certain disease with a normal gene, thus preventing the patient from getting the disease in the first place.

[*See also* **Cell; Double Helix; Gene; Nucleic Acid**]

Dominant and Recessive Traits

Dominant and recessive traits exist when a trait has two different forms at the gene level. The trait that first appears or is visibly expressed in the organism is called the dominant trait. The trait that is present at the gene level but is masked and does not show itself in the organism is called the recessive trait.

In order to understand the concept of dominant and recessive traits, it is necessary to know what is meant by the word "allele." Alleles have to do with genes, and genes are the carriers of information that determine an organism's traits. Our height, hair color, blood type and overall looks are but a few examples of traits that are the result of the chemical activities directed by our genes. Every human being is produced by sexual reproduction and therefore receives twenty-three gene-containing chromosomes (coiled structures in the nucleus of a cell that carries the cell's DNA) from each parent, resulting in a full complement of forty-six chromosomes. When the chromosomes pair up to form a new and unique individual (since chromosomes always exist in pairs), they do so in a very particular way so that the same trait is always carried on the same place or position on the chromosome. In other words, since the offspring receives information on each trait from both its parents, there are corresponding pairs (or two genes) that match together for each trait. Sometimes these are the same (when a person inherits a gene for blood type O from both its parents), and sometimes these are different (when the person inherits blood type O from the father and blood type A from the mother). When these forms of the same type of gene are different or alternative versions, they are called "alleles." Therefore, alleles are different forms of a gene for a particular characteristic. However, more and more the word allele is used interchangeably for gene.

Most often when an individual receives two different alleles for a given trait, one allele is expressed and the other is not. For example, a person may receive one allele for a straight hairline and another for a widow's peak (when the hair comes down to a point in the middle of the forehead). In such a case, the person will have a widow's peak since that allele is "dominant" or is the one that is able to express or show itself. In the same way, the allele for brown eyes is known to be dominant over the allele for blue eyes. Conversely, the allele that is masked or is not able to show itself (despite being there) is called "recessive."

Austrian monk Gregor Mendel (1822–1884) made the first detailed investigation of inherited traits in the 1860s. Since Mendel's groundbreaking work, the rule has been that when two organisms showing different traits are crossed, the trait that shows up in the first generation is considered the dominant trait. A dominant trait could be compared to an athlete who dominates a game or a person who dominates a conversation. Each of these people monopolize things to the point where others have no chance to express their ability or ideas. It is in this way that a dominant allele expresses itself and suppresses or masks the activity of the other allele for that trait. Although the masked allele is not expressed, it is still there and remains part of the person's inherited package. This means the recessive allele can still be passed on to the next generation. Masked or recessive traits can only express themselves when the individual has a matching recessive allele (totaling two alleles for that trait).

Although Mendel did not know exactly what the gene and the allele were, he knew very well that they existed in some form (he called then "factors"), and that they followed certain laws. He was therefore able to formulate what became known as the law of dominance. This law states that when a dominant and a recessive form of a gene come together, the dominant form masks the recessive form. Thus, even though the recessive allele (or member of the gene pair) is still present, it is not visible.

[*See also* **Gene; Inherited Traits; Mendelian Laws of Inheritance**]

Double Helix

The double helix refers to the "spiral staircase" shape or structure of the deoxyribonucleic acid (DNA) molecule. DNA is the genetic material of all living organisms. Also described as a twisting ladder, this double helix model enabled scientists to finally account for both the similarities as well as the immense variety of life.

Double Helix

The discovery of the double helix molecular structure of DNA in 1953 by American biochemist James Dewey Watson and English biochemist Francis Harry Compton Crick was one of the major scientific events of the twentieth century and some would say in the history of the life sciences. Prior to this discovery, it was not understood how such a relatively simple nucleic acid as DNA could contain such a vast complex of hereditary material, and few, if any, believed that it did.

As a nucleic acid, DNA was known to be composed of only four different submolecules called nucleotides—adenine (A), guanine (G), thymine (T), and cytosine (C)—and Watson and Crick believed that if they could determine the structure of DNA, they could explain how DNA actually works. The major way of learning about the structure of a chemical is to crystallize it and x-ray the crystals. When x rays pass through the crystals, they bend or diffract and create a pattern that can be studied. Watson and Crick worked with English physicist Maurice H. F. Wilkins and his associate Rosalind Franklin, whose excellent x rays in

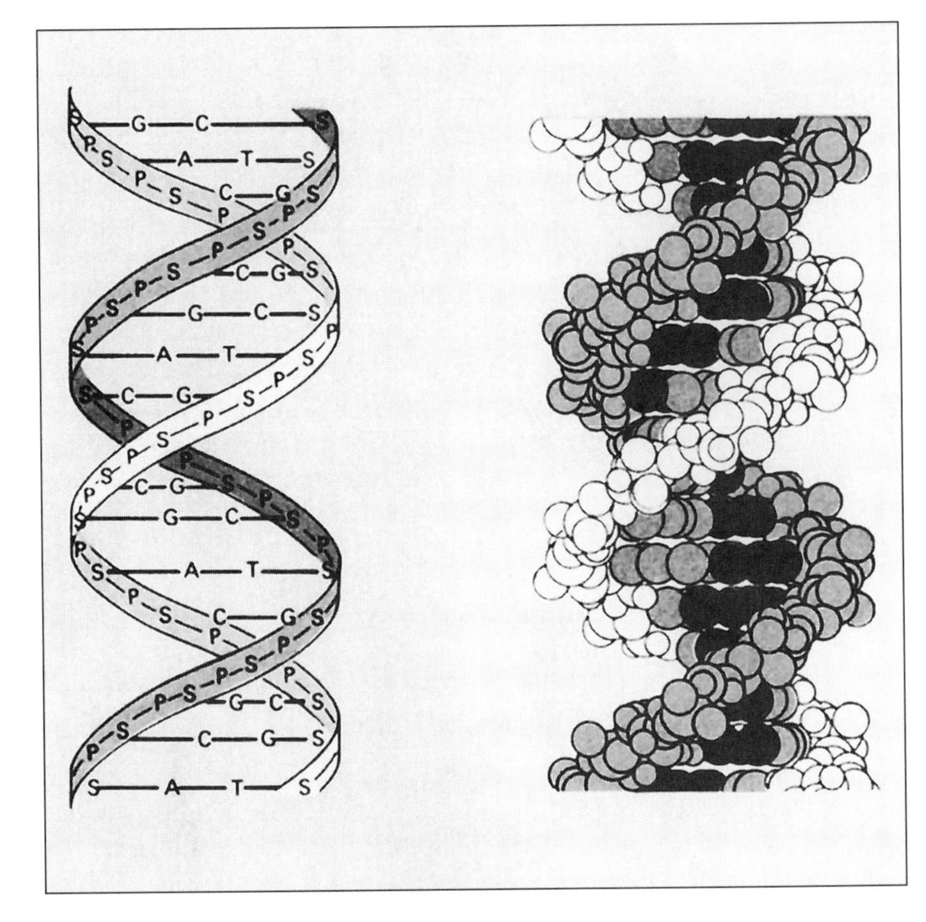

The structure of a DNA molecule. Watson and Crick knew that the structure of DNA held the key to understanding how genetic information is stored and transmitted. (Reproduced by permission of the Federal Aviation Administration. Photograph by Robert L. Wolke.)

1951 provided important evidence that DNA had a spiral shape. Franklin's x rays showed that the DNA molecule was a double strand of twisted material which came to be called a double helix. (Helix is taken from the Greek word for spiral.)

Knowing this did not allow Watson and Crick to correctly describe the actual structure of a DNA molecule, and for some time they tried to build a model of what it might look like. Although, until they discovered that the four nucleotides always formed themselves into a definite pattern of pairs (A always pairs with T, and G always pairs with C), they were unable to make further progress. Once they knew that these "base pairs" were complementary (in other words they always paired up the same way), Watson and Crick designed and built a model in which the correctly paired bases were the "rungs" of a ladder that connected the two sides or "rails" of the ladder. These sides were then twisted in the shape of a compact spiral or coil. Watson and Crick also explained that the rungs on the ladder (called bases) were the coded instructions, and that the order of these four nucleotides (A,T,G,C) spelled out the instructions for all of the different characteristics of an organism.

In March 1953, the two scientists announced their discovery of the double helix structure of the DNA molecule, offering to science what was basically the explanation of the chemical basis of life itself. Unlike many discoveries in the life sciences, knowledge of the structure or shape (the double helix) of the DNA molecule was essential to explaining how it could carry all the information needed to make a living creature, as well as how it could make exact duplicates of itself. In 1962 Wilkins shared the Nobel Prize in Physiology or Medicine with Watson and Crick for their discovery. Rosalind Franklin would surely have been included as well, but she had died in 1958 and the prize is only given to living scientists.

[*See also* **DNA**]

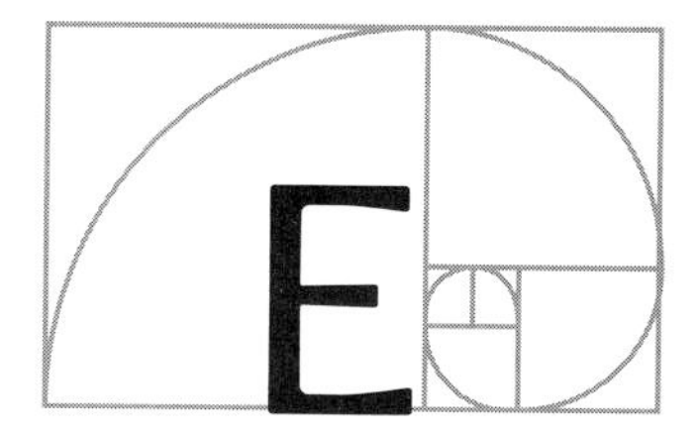

Echinoderm

An echinoderm is a spiny-skinned invertebrate (an animal without a backbone) that lives in the ocean. Most echinoderms, like starfish and sand dollars, have a distinctive five-part body plan, an endoskeleton (an internal skeleton), and many tiny, sucker-tipped tube feet with which they take in seawater in order to move.

As one of the more unusual animals, the echinoderm (meaning "spiny-skinned") numbers about 6,000 related species and includes such exotic animals as the five-armed starfish, the pincushion-like sea urchin, and the bottle-shaped sea squirt. With their armor-like external spines that protect them, it might seem that an echinoderm has an exoskeleton (like a lobster or an insect). In fact, the spines are only extensions of an interior or endoskeleton that makes echinoderms unique among invertebrates. This internal skeleton is made of plates under the skin that have spiny projections. Because of this internal skeleton, echinoderms are considered closer to vertebrates that any other invertebrate phylum. However, unlike vertebrates, echinoderms have no head or centralized nervous system. Nor do they have an excretory or respiratory system. Finally, echinoderms have a circulatory system that is unique in the animal kingdom since they use seawater as their circulatory fluid. Echinoderms use their specialized tube feet to suck in water and create a suction effect that allows them to move about and to feed on other animals.

The starfish is an excellent example of features that make an echinoderm such a different animal. Also called sea stars, this invertebrate has a radial symmetry, meaning that its body parts are arranged around a central area or hub. Animals with radial symmetry can be divided into mirror-image halves along many lengthwise lines. A human body that has bilateral symmetry only can be divided perfectly down the middle and, therefore, have only two mirror-image halves. However, a starfish with five arms can be divided evenly several different ways. A starfish has a skeleton made up of hard plates beneath its skin. It also has many canals inside its body that help it operate a water pumping system, allowing it to move about and find the food it needs. These canals are connected to tube feet that are located on its underside and are hollow, suction-cup-like structures that suck in water. This forces the tube feet outward. A starfish can walk on the tips of its arms, using the sucker action of the tube feet to cling to a surface. When the suction is released, it moves forward. The starfish also uses its tube feet to wrap its arms around a clam shell and pry its hinged shells apart. A starfish's mouth is located in the center of its underside. It eats the soft-bodied mollusk by pushing part of its own stomach through its mouth and engulfing its prey, thereby digesting the clam inside the shell.

A starfish on the California coast. The starfish is an excellent example of an echinoderm because of its five-pointed, radially symmetrical arms. (Reproduced by permission of Field Mark Publication. Photograph by Robert J. Huffman.)

Finally, the starfish and all other echinoderms reproduce sexually by the union of male sperm and female egg. They also have remarkable powers of regeneration. If an echinoderm's arm breaks off, it will grow back, and if a part of its central core breaks off, it will grow into a complete organism. Many clam and oyster fishermen viewed the starfish as a rival, since clams and oysters are the starfish's main diet, and would tear these creatures apart whenever they found them. Unfortunately the fishermen were only multiplying their problem.

[*See also* **Animals; Invertebrates**]

Ecology

Ecology is the branch of the life sciences that studies the relationships between living things and their environment. The basic principle behind the idea of ecology is that organisms interact with their surroundings and are influenced by them, while at the same time changing them. The notion of ecology is therefore something like a two-way street. The study of ecology takes as a basic principle something that we all recognize once it is stated: the dynamic, living world is based on complex relationships between organisms and the places in which they live. Therefore, ecology can be described as the attempt to explain how and why living things interact with each other and their environments.

The word ecology comes from the Greek words *oikos,* meaning "house, household, or place to live" and *logos,* meaning "the study of." The word was created in 1866 by Ernst Heinrich Haeckel (1834–1919), a German biologist and philosopher, after he came to recognize the importance of studying the environment as a separate scientific field. Haeckel was greatly influenced by the English naturalist Charles Darwin (1809–1882), whose classic book, *On the Origin of Species by Means of Natural Selection,* inspired Haeckel to think about the important role the environment played in the story of nature. Until Haeckel, ecological issues were referred to as natural history, and a student of nature was called a "naturalist." While there was some notion of the interrelatedness of life and its environment before Darwin's work, it was mostly a notion of nature as being economical and Earth being a place where things were always balanced harmoniously.

ECOLOGY DRAWS ON MANY SCIENCES

Today, scientists who study ecology are called ecologists, and they are well aware that although there may be a type of harmony in nature,

RACHEL LOUISE CARSON

American biologist Rachel Carson's book, *Silent Spring*, is one of the most important works in the history of ecology (the study of the interrelationships of organisms and their environment). This very popular book became an instant classic and informed people about the severe side effects of chemical poisons on all forms of life, thus making people aware for the first time that environmental damage can affect life itself. Her work marks the beginning of the environmental movement in the United States.

Rachel Carson (1907–1967) was born in Springdale, Pennsylvania. Her interest in nature came from her mother taking her into the woods and fields around her home and exploring the countryside. This love of nature would remain with Carson throughout her life, and after entering the Pennsylvania College for Women as an English major she was so impressed by a biology teacher that she switched her major. In 1929 she graduated with a degree in zoology and went on to earn her Master's degree in zoology from Johns Hopkins University in 1932. Her appointment to the Department of Zoology at the University of Maryland allowed her to teach and spend summers doing research at the Woods Hole Marine Biological Laboratory in Massachusetts. This turned her interest more to the sea, but family commitments to her widowed mother and orphaned nieces soon forced her to leave teaching and take a position at the United States Bureau of Fisheries.

It was while she worked at the bureau that she began to combine her love of nature with her love of writing. In 1941 she produced her first book about the sea, and during World War II (1939–45) she wrote fisheries information bulletins for the government. By 1949 she had become chief biologist and editor for the new United States Fish and Wildlife Service, and began a new book titled *The Sea Around Us.* This book appeared in 1951 and won the National Book Award, becoming an immediate best-seller. After a second successful book on the sea, Carson produced her classic, *Silent Spring,* in 1962. This book has been described as the work that gave rise to the environ-

the balance supporting it is both fragile and difficult to repair if broken. If ecologists can be said to be specialists, their specialty is necessarily a very general one. This is the case because ecologists use knowledge from many disciplines, such as chemistry, computer science, mathematics, and physics, besides the many subspecialties of biology (botany, microbiology, morphology, physiology, and zoology). This multidisciplinary approach is required since ecologists' subject matter is so broad and deals with all major living and nonliving aspects of what can be called a system. Increasingly however, there are more individuals who are becoming specialists within the general field of ecology, resulting in such spe-

mental movement in the United States, and Carson was the first to draw widespread attention to the fact that the haphazard use of chemical pesticides was creating an ecological disaster for animals and people alike. Carson was above all a scientist, and her book was not a one-sided argument full of accusations and no facts. Rather, it was a well-documented, reasoned analysis of the actual and potential hazards of misusing chemicals.

Carson's book focused specifically on the disastrous effects caused by the highly popular but overused pesticide DDT (declare-diphenyl-trichloroethane). Carson showed that birds would eat insects contaminated with DDT, and as a result lay eggs with shells too brittle for the birds inside the egg to survive to birth. She also documented how DDT would remain poisonous for years since it was unaffected by the sun, rain, or even bacteria or acids in the soil. It would also be stored in body tissues after being eaten and therefore would remain in the food chain (the series of stages energy goes through in the form of food), killing more birds, fish, and small animals.

Despite extreme criticism and attacks by the chemical industry, Carson's book served to awaken people that they were poisoning their own environment. It led to the creation of a Presidential Advisory Committee on the use of pesticides and had a great deal to do with the eventual creation of the Environmental Protection Agency. Carson's book is rightly considered to be a landmark in the environmental movement since it did more than just alert people to the dangers of pesticides. Perhaps even more important, it took a holistic, or overall view, of life on Earth and showed how all living things were connected and how they ultimately affected one another. This led directly to an awareness of the meaning and importance of ecology. Carson was therefore a pioneering environmentalist and ecologist whose work and dedication inspired a new worldwide movement of environmental concern. Her work will remain a hallmark in the increasing awareness that all people have about how humans interact with and affect the environment in which they live and on which they depend.

cialists as population ecologists and restoration ecologists, to name just a couple.

ECOLOGY, THE TWENTIETH-CENTURY SCIENCE

Although ecology began with Haeckel in the 1860s, it is really a twentieth-century science, and it is therefore a relatively young one. As a science, ecology is founded on four basic principles upon which most ecologists agree. First, it states that an organism's life pattern or the way it lives reflects the patterns of its physical environment. This is true for nearly all plants and animals who survive and prosper when certain favorable conditions are present (unlike humans who can manipulate their surroundings and create a livable place). The second principle states that living creatures in a certain place tend to group themselves into loosely organized units known as "communities." A community exists wherever a species has its natural home. The third principle of ecology states that an orderly, predictable sequence of development (change) takes place in any area. The slow process of orderly change over time is called "ecological succession," and is more understandable once we realize how interdependent all life is with other life forms and its environment. The fourth principle states that a community and its environment make up an ecosystem. The notion of an ecosystem is important because by dividing the natural world into ecosystems it becomes easier to study. Since an ecosystem describes any system that has organisms interacting with other living things as well as their environment, an ecosystem can be anything from a fallen tree rotting in the forest to something as big as a river.

POPULATIONS, COMMUNITIES, AND ECOSYSTEMS

In studying the relationships between living things and their environment, ecologists regularly use three main organizing ideas that have proved very helpful. These are the concepts of populations, communities, and ecosystems, which allow ecologists to categorize systems in terms of how they are organized. A population is the smallest system and is defined as a group of individuals from the same species who live in the same area at the same time. For example, the bees in a hive represent a population, but so do a particular forest of beech trees or the deer that live in that forest. Studying a particular population over time allows an ecologist to know what natural forces can change the size of a population (like bad weather or hunting by predators), and to detect when human activities are at fault (such as sewage runoff or acid rain).

A community is larger than a population and is defined as a group or a collection of different animal and plant populations that live in the same area. It is also sometimes called a "biotic community" since it is made up of living things (biotic means living things). Ecologists also study the roles that different species play in their communities. They call this the "ecological niche" of a species, and it refers to such factors as what it an animal eats or what amount of moisture is necessary for a certain plant to thrive. More broadly, a niche has been described as the exact way in which a living thing fits into its environment. An example of a highly specialized niche would be an animal that is the only species that eats a certain plant to live. The animal's specialization serves it well since it has no competition for its food. However, this specialization could be dangerous as well, since the animal might starve if the particular plant suddenly dies off. Finally, an ecosystem (or an ecological system) includes a community (biotic) as well as a community's environment (abiotic).

LANDSCAPE, BIOME, AND BIOSPHERE

Besides population, community, and ecosystem, ecologists may use larger categories to describe the environment that makes up an ecosystem. Beginning with a particular ecosystem, the next largest unit is a landscape that includes groups of ecosystems and humans. Landscapes are part of a bigger unit called a biome. Biomes are large geographical areas that include certain type of climates, vegetation, and animals. The six major land biomes are rain forest, desert, grasslands, temperate deciduous forest, taiga, and tundra. Larger than biomes are biogeographic regions such as major continents and oceans. The final and largest category of all is the biosphere. This includes every part of Earth that has living things on it, from the sky to the ocean bottom to the mountain tops. The biosphere includes all of Earth's ecosystems functioning together. The idea of functioning together is especially important, since the notion of every part of the whole being linked together is what the concept of ecology is all about.

DIVERSITY CREATES HEALTH

Knowing how living and nonliving things are linked together helps ecologists understand better how living things depend on each other for survival. This knowledge allows certain types of ecologists called "restoration ecologists" to work toward returning a damaged ecosystem back to its natural state. More recently, one of the areas receiving attention from ecologists is the notion of the importance of biodiversity to the environment. Only in the 1980s did scientists begin to realize that the more diverse and complex an ecosystem was the healthier it was. It was also realized that the loss of biodiversity can be permanent, as when a species goes extinct (life forms that have died out). Since ecosystems interact with one another in ways that we sometimes do not understand, human activity could be harming biodiversity and permanently altering many of Earth's ecosystems. Ecologists are also concerned about many other ways that human beings affect the planet, and many feel that if we continue to pollute our air and water, destroy the rain forests, and steadily increase our population, our harm to Earth may be irreversible.

[*See also* **Biome; Biosphere; Community; Ecosystem; Population**]

Ecosystem

An ecosystem (ecological system) is a living community and its nonliving environment. It is a term used by life scientists to break up the biosphere (the entire living world) into smaller parts so as to more easily categorize and study those parts.

An ecosystem can be any size and has no set boundaries. The term can be applied to an entire forest, a lake, a vacant city lot, a suburban lawn, and even a crack in a sidewalk. An ecosystem is a complex system made up of communities of living organisms that interact with each other and with their nonliving surroundings. Ultimately, the entire Earth with all of its life and its physical environment can be said to make up the largest ecosystem of all—Earth's biosphere.

BIOTIC AND ABIOTIC COMPONENTS OF AN ECOSYSTEM

All ecosystems, no matter their size, have two interacting parts—the biotic (living) component and the abiotic (nonliving) component. The biotic component is made up of autotrophic organisms (self-nourishing) and heterotrophic (other-nourishing) organisms. Green plants are autotrophic since they can make their own food or nourish themselves by their ability to convert sunlight into energy. Animals are examples of heterotrophic organisms since they cannot make their own food and are able to break down other living things (plants or animals) and use their energy.

OTHER PARTS OF AN ECOSYSTEM

These biotic and abiotic components are in fact only two parts of a six-part system that ecologists use to categorize what goes on in every ecosystem. These six elements are based on the flow of energy and the cycle of nutrients within an ecosystem. These six elements are (1) the sun; (2) abiotic substances; (3) primary producers; (4) primary consumers; (5) secondary consumers; (6) decomposers. The sun of course is the ultimate source of all energy, and its light is used by green plants (as primary producers) to make food in a process called photosynthesis. Essential to this process are abiotic substances like carbon dioxide, water, and phosphorus, which the plant uses to carry out its food-making series of chemical reactions. After this, a green plant is ready to be consumed or eaten by a primary consumer called a herbivore (any plant-eating animal from a mouse to a cow or an insect) or an omnivore (any animal able to eat both plants and other animals). Primary consumers are followed, sometimes literally, by secondary consumers who are carnivores and therefore

FERTILIZATION OF EGGS

Eggs are usually fertilized in one of two ways. In mammals, snakes, birds, turtles, and insects, mating or the coming together of male and female sex cells takes place inside the body of the female. Unlike mammals, most aquatic animals, like fishes, toads, and frogs, engage in external fertilization. For example, the male frog sprays the female's eggs with his sperm as she deposits them in the water. For many animals, including humans, eggs (that are laid or spawned) are a source of food since they contain a great deal of nourishment.

[*See also* **Fertilization; Reproduction, Sexual; Reproduction System**]

Embryo

An embryo is an early stage in the development of an organism. It can be applied to plants as well as animals and is the most important growing phase of an organism. The embryonic stages of all animals are similar, although they may occur at different times.

All complex organisms, from a plant to a human being, begin as a single fertilized cell. This cell is called a zygote and it is the product of a sperm (a male sex cell) and an egg (a female sex cell) coming together. From the moment this fertilized egg or ovum begins to divide until it has formed its organs, it is known as an embryo. In some animals, like birds, this takes place inside an egg that is expelled from the female's body. In other cases, like most mammals, this takes place inside the body of the female. In humans, the word embryo refers to the first two months of development, after which it is called a fetus.

In the development of every organism from a single cell to a complex living thing, a great many stages or phases take place. The embryonic stage is one of these, and it also has phases or stages of its own. This embryonic stage is probably the most important of all the stages because it is during this time that the growing embryo forms all of the tissues and structures that it will need as a mature adult.

FROM EMBRYO TO ORGANISM

A developmental chain begins within an embryo that eventually results in the birth of a fully developed organism. The first stage is called the embryonic stage, and it has four of its own stages: cleavage, gastrulation, neurulation, and organogenesis. Beginning with the zygote, the first stage of cleavage or repeated cell division occurs when the zygote cleaves or splits into two cells. This division continues as two cells become four, four become eight, and so on, until a ball-shaped cluster is produced called a morula. As division continues, the morula becomes a blastula, which also resembles a ball of cells. It is at this point in mammals that attachment to the uterus occurs. Soon the blastula folds inward and gastrulation, or rearrangement, of the cells begins. This gastrulation creates layers of cells, and in vertebrates each layer will produce different organs. This begins the process of specialization known as differentiation.

As differentiation continues, the first specialized tissues begin to form, starting the neurulation phase. During neurulation, organs form at the head of the new organism and continue to develop downwards. By the time organogenesis begins, the embryo has also started to change its shape to make its developing organs fit. The body becomes longer, and a head and trunk become visible. Dramatic and rapid changes now occur as the em-

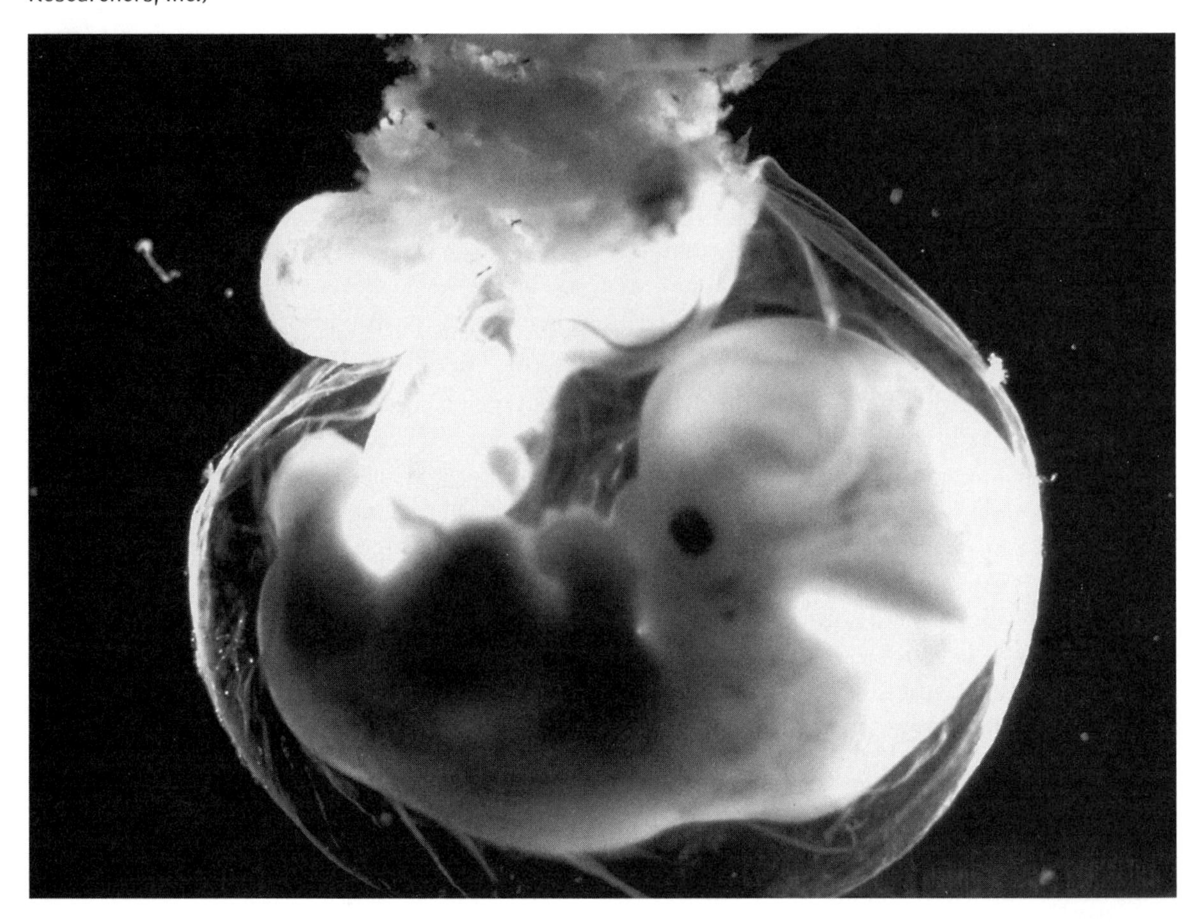

A human embryo at five to six weeks of development. (Reproduced by permission of Photo Researchers, Inc.)

bryo develops rudimentary or primitive versions of the heart, lungs, liver, kidneys, and other vital organs.

Although all animals follow the same stages, their time schedules can be very different. While a fish or an amphibian may be ready for birth after six weeks, mammals, like humans, must remain in the female's womb for another twenty or so weeks to develop further. However, by the time the human embryo reaches eight weeks, all its adult organ systems are in place and it is no longer called an embryo, but is a fetus. Plants also develop from embryos. In flowering plants and conifers (like evergreens), the embryo is the part of the seed from which the mature plant develops. In plants like ferns (nonflowering plants that form spores), the embryo is the mass of cells that develops into a new plant.

[*See also* **Fertilization; Plant Anatomy; Plant Reproduction Reproduction, Sexual; Zygote**]

Endangered Species

An endangered species is any species of plant or animal that is threatened with extinction. Although extinction, or the situation in which no living member of a species exists any longer, can have natural causes, it is often being hastened by the activities of humans. International efforts to identify the most threatened species have begun, and various national and international laws have been passed to protect these endangered organisms.

Biologists estimate that about 500,000,000 species of plants and animals have existed since life began on Earth. Since there are only some 2,000,000 to 4,000,000 species in existence today, most of the species that ever lived on Earth are now extinct. This demonstrates that extinction is a natural phenomenon that occurs as the normal process in the course of evolution. Evolution has shown that species that cannot adapt to natural changes in their environment or to increased competition from other species will eventually become extinct. Besides climate or environment change and competition, disease and natural catastrophes can destroy an entire species. It is believed that such an occurrence took place more than 65,000,000 years ago when an asteroid struck the Earth and eliminated more than half of the planet's plant and animal life, including the dinosaurs.

ENDANGERMENT INCREASES DUE TO HUMANS

Although extinction may occur naturally, the growing problem in our modern world is that increased human activities have dramatically increased the natural rate of extinction. While people have always had an effect on their environment, the degree to which they have affected it has increased substantially since the Industrial Revolution of the eighteenth century. During the second half of the twentieth century in particular, technological advances combined with a rapidly expanding human population have changed the natural world as never before. As the twenty-first century begins, biologists estimate that the rate of extinction has increased to somewhere between 100 to 1,000 times nature's normal rate. Today, species are threatened and become endangered primarily because of a combination of three human factors: their habitat is either disturbed or eliminated; they are overhunted; or they are being eliminated by other nonnative "introduced" species.

CAUSES FOR ENDANGERMENT

Habitat destruction is the main reason for the increasing number of today's endangered species. The steadily increasing human population has in many ways taken over areas that formerly were habitats or homes for certain organisms. This "takeover" usually shows itself in the form of houses, highways, and industrial buildings in developed countries, and growing farms and farmland in less developed ones. Whenever a new airport or dam is built in a formerly natural landscape, the life forms that lived there must somehow either adapt, move, or die. Often a habitat is totally destroyed and changed into something that is totally unrecognizable from its former self. Other times however, habitat destruction is less obvious, such as when industrial runoff or city sewage degrades the quality of a habitat but leaves it apparently looking the same. As habitats are regularly chipped away and become smaller and smaller, the pressure on the organisms in the habitat becomes greater, and it becomes harder for them to survive.

The inhabitants in a given habitat are often also sought after by people for many different reasons. Whether people have killed animals for sport (such as the American bison was in the nineteenth century) or for profit (like the whale and the rhinoceros), overhunting has placed many of today's species close to extinction.

Finally, people have altered habitats dramatically by taking one species from its native habitat and transplanting it into a different habitat somewhere else. Sometimes this is done deliberately, as when rabbits were introduced to Victoria, Australia, by Thomas Austin in 1859. Austin release twenty-four of the animals from England to be used for sport hunting. However, within twenty years these twenty-four rabbits multiplied to millions. Not only did they became serious pests, but they set about actively destroying certain vegetative life.

Other times, introducing a new species is done accidentally, as when the gypsy moth was accidentally introduced into the United States from France in 1869. Sometimes newly introduced species cause little or no harm, but many times they upset the delicate balance that was established by the native organisms, and often out-compete these "natives" for scarce resources. Sometimes a new species can actually eliminate a species that was perfectly adapted to its habitat but had no natural defenses to use against the newcomer.

SAVING ENDANGERED SPECIES

Fortunately, something is being done for species that are so threatened that they may disappear altogether. On a global level, the International Union for Conservation of Nature and Natural Resources (IUCN) classifies such species as endangered, critically endangered, threatened, or rare. To the IUCN, an endangered species is in the greatest danger and faces immediate extinction, possibly even if action is taken in an effort to prevent it. A critically endangered species is one that will not survive without human help. Threatened species may still be abundant in their own habitat, but overall their population is rapidly declining. Rare species are considered at risk because of low overall population numbers. The Endangered Species Act of 1973, which became law in the United States that year, obligates the government to protect all animal and plant life threatened with extinction. It also provides for the drawing up of lists of such species and promotes the protection of critical habitats or places that are essential to the survival of a species. One of the first species placed on that list was the nation's own symbol, the bald eagle. In May 1998, the bird was taken off the list, along with several other species that have apparently been saved.

[*See also* **Extinction; Habitat**]

The pygmy lemur of West Madagascar is the world's smallest primate and in danger of becoming extinct. Human actions have led to the endangerment of this species. (Reproduced by permission of World Wide Fund For Nature Photolibrary.)

Endocrine System

The endocrine system is made up of a number of glands that control bodily functions by secreting hormones. It is found in all vertebrates (animals with a backbone) and works together with the nervous system to regulate many of the body's functions. The hormones secreted by glands are chemicals that carry instructions from one set of cells in an animal's body to another. The endocrine system can be described as the body's chemical coordinating system.

Although the endocrine system works closely with the body's nervous system in helping it respond appropriately to a changing environment, it is different from that system in many ways. Where nerve transmission is rapid and over quickly, the effect generated by a certain hormone can last several days and is usually concerned with large, long-term processes or situations. For example, the nervous system makes a child jerk his hand away from a hot stove almost as soon as he or she has touched it. In a very different situation however, it is the endocrine system that senses that the child has not been fed for some time and makes sure that his or her brain has the steady supply of glucose (sugar) that he or she needs. Therefore, the messages sent by the endocrine system usually have long-term effects and are almost always concerned with the body's larger processes, unlike the nervous system which generally brings about short-term changes.

THE ENDOCRINE SYSTEM AND HORMONES

In vertebrates, the endocrine system uses several different glands, as well as specialized tissue and cells, to secrete hormones directly into the bloodstream. These glands are called ductless because they have no need for ducts or tubelike connections to the circulatory system (a network that carries blood throughout the body). Rather, they secrete hormones in very small amounts that travel throughout the body via the bloodstream to eventually reach their "target organ" (which is sometimes not close to the secreting gland). Although the hormones cause no changes in the tissue they pass through, they do cause major changes in the target organs, which have special receptors or receivers for hormones. Working together with the nervous system, the hormones released by the body's endocrine system regulate growth, ovulation (the release of an ovum or egg from an ovary), milk production, sexual development, and many other processes. The endocrine system can be considered the body's chemical coordinating system as it gives orders to start and stop certain bodily changes as needed. Since stopping a process is sometimes as important as starting, the endocrine system has what are called feedback mechanisms that monitor a situation and signal when a gland should stop secreting hormones. Malfunctions in the system can cause an over or undersupply of a hormone, resulting in such conditions as gigantism (unusually large stature), dwarfism (unusually small stature), or a goiter (visible swelling at the front of the neck).

Hormones also play a large part in the growth cycle of insects. Since insects have an exoskeleton (a hard shell that surrounds its body), they must periodically molt or shed their skeleton if they are to grow. Molting involves the precise coordination of many separate steps, and hormones begin and coordinate the entire, complicated process. A hormone also signals caterpillars to make a cocoon and enter their pupa (resting) stage. It is in this stage that they undergo a metamorphosis and transform into an entirely different creature, like a moth or butterfly.

GLANDS OF THE ENDOCRINE SYSTEM

In humans and other vertebrates, the endocrine system regulates a great deal more than one or two bodily functions and processes. In humans, there are at least nine separate glands that make up the endocrine system, and they are scattered throughout the body. These major endocrine glands are: the pituitary, the hypothalamus, the pineal, the thyroid, the parathyroid, the thymus, the adrenal, the pancreas, and the gonads.

The Pituitary. Located at the base of the brain (of which it is actually a part), the pituitary influences so many glands that it has been called the "master gland." About the size of a pea, this gland works with the hypothalamus as part of a direct link between the endocrine system and the nervous system. Although small, it is really two glands in one and its parts are called the anterior (front) and the posterior (rear) pituitary. The anterior secretes the hormone prolactin, which makes the female body ready to produce milk. The anterior also secretes five other hormones that start other glands working. The posterior pituitary secretes oxytocin, which makes the uterus contract during birth and stimulates the release of milk. Vasopressin which regulates the balance of water and raises blood pressure is also released by the posterior pituitary.

The Hypothalamus. The hypothalamus monitors internal organs and emotional states and supervises the release of hormones from the anterior pituitary gland. It also produces substances called releasing factors that control hormonal secretions from other glands.

The Pineal. The pineal gland is a light-sensitive organ that evolved from a third eye that vertebrates had on top of their heads until about

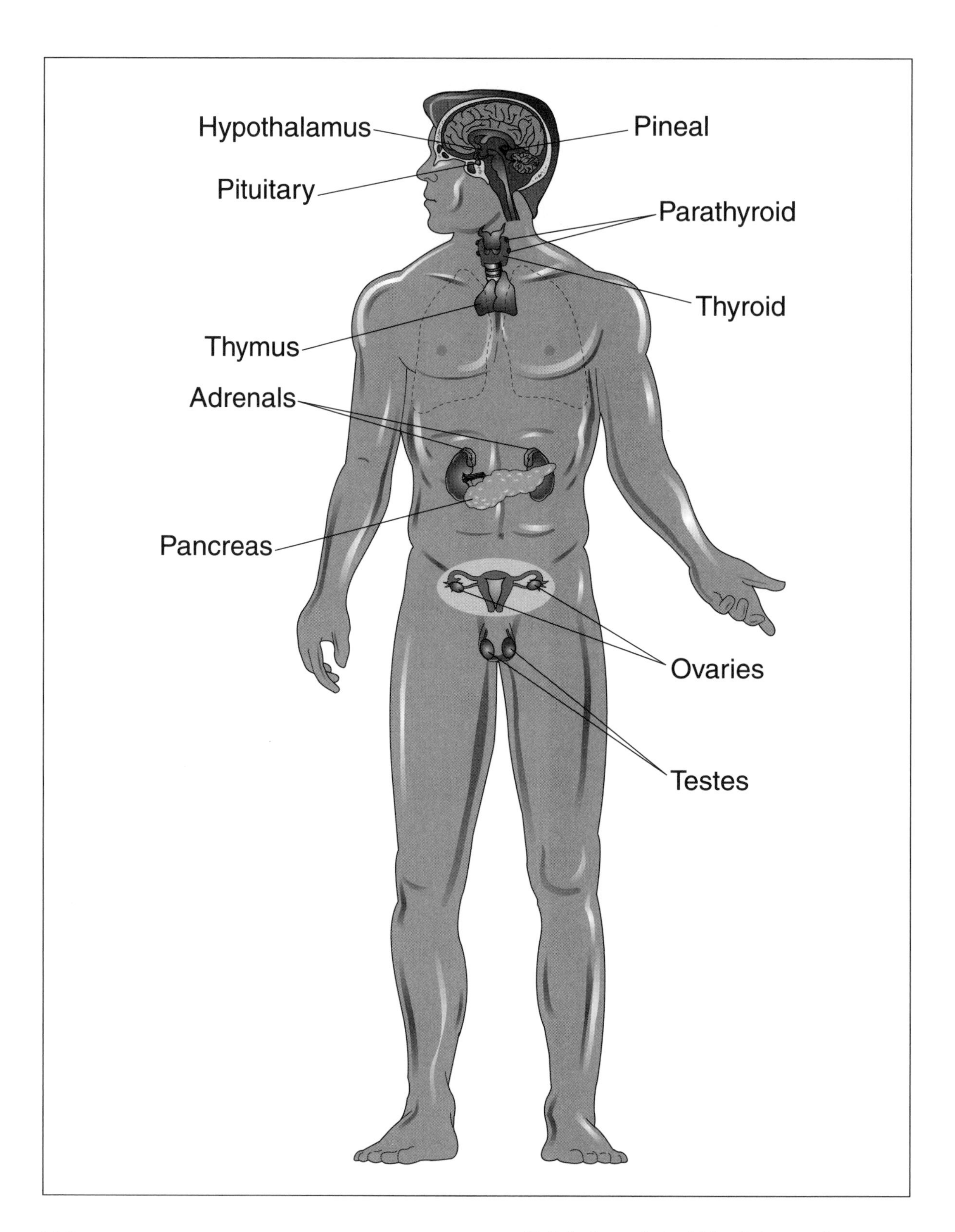
Hypothalamus
Pineal
Pituitary
Parathyroid
Thyroid
Thymus
Adrenals
Pancreas
Ovaries
Testes

240,000,000 years ago. It secretes the hormone melatonin, which acts as the body's biological clock and helps regulate its rhythms.

The Thyroid. The thyroid is located around the windpipe at the front of the neck and produces hormones containing iodine that are necessary for growth, development, and proper metabolism (all the body's processes involved in using energy). Hypothyroidism, or a lack of these hormones, leads to stunted growth and retardation in children. In adults, hypothyroidism produces dry skin, sluggish behavior, and an inability to tolerate cold. Hyperthyroidism, or an overproduction of hormones, results in nervousness and agitation, weight loss, and heavy sweating.

The Parathyroid. There are four parathyroid glands next to the thyroid. They regulate the level of calcium in the blood by taking calcium stored in bones and releasing it into the bloodstream. Proper calcium levels are essential to muscle contraction and the clotting of blood.

The Thymus. The thymus is located behind the breastbone and between the lungs. It is often considered to be part of the immune system since it plays a major role in the development of lymphocytes (white blood cells) that fight infection. It is also considered part of the endocrine system because it produces a hormone that stimulates the growth of these white cells.

The Adrenals. The two adrenal glands found in humans are located above each kidney, and their outer region, called the cortex, secretes hormones called glucocorticoids. These help maintain the blood's important glucose level and also minimize inflammation (swelling and tenderness) that is caused by infection or injury. The adrenal medulla is the inner region of the gland and secretes epinephrine and norepinephrine, which prepare the body for stress by narrowing the blood vessels and increasing both the heart rate and the amount of glucose in the blood. During times of major excitement, these dramatic changes can be easily felt and recognized as the body prepares for "fight or flight."

The Pancreas. The pancreas (an organ involved in digestion) is located below the stomach and contains clusters of endocrine cells called pancreatic islets that secrete glucagon and insulin. These hormones are key to controlling the level of glucose in the bloodstream. When not enough insulin is produced or the target cells do not respond to insulin, a disease called diabetes mellitus is diagnosed.

The Gonads. Finally, the gonads are the main organs of reproduction and secrete sex hormones. The testes in males and the ovaries in females usually come in pairs in most mammals. The testes are responsible for producing androgens—the most important of which is testosterone—

Opposite: A labeled illustration showing all of the glands of the human endocrine system. All of the glands are same for males and females except for the gonads. The ovaries are found in females and the testes are found in males. (Illustration by Kopp Illustration.)

which help develop sperm and are responsible for what are called secondary sexual characteristics, like a deeper voice and facial hair. The ovaries help in the production of ova (eggs cells) and produce progesterone and estrogen. These hormones are responsible for female development (like breasts) and also control pregnancy. Both of these physical changes come about in boys and girls during puberty, that burst of development that occurs during the early teenage years.

THE IMPORTANCE OF THE ENDOCRINE SYSTEM

It is thought that vertebrates have as many as fifty different hormones, suggesting that the endocrine system of humans is a highly complex and very intricate system. It might be said that the overall goal or purpose of the endocrine system is to maintain a balance in all the body's systems. Despite this general goal, the hormones that the system actually releases can have powerful and even dramatic effects. For example, the gonads change a boy into a man and a girl into a woman; the adrenal gland pumps adrenalin and can make a person unnaturally strong or not feel pain; the pituitary can produce nutritious breast milk to nourish a baby; and the thymus can help wage and win a war against infection.

[*See also* **Hormone; Immune System; Reproductive System; Sex Hormones**]

Endoplasmic Reticulum

Endoplasmic reticulum is a network of membranes or tubes in a cell through which materials move. As an organelle in a eukaryotic cell (complex cells having nuclei and other organelles), it is involved with the production of new proteins as well as with the movement of materials throughout the cell.

The name endoplasmic reticulum means "an inside formed network," and these words properly describe this network of connected folds, tubes, and sacs that are found in all eukaryotic cells (those with a distinct nucleus). As an organelle in a cell, the endoplasmic reticulum is a specialized structure and has certain tasks to perform. The endoplasmic reticulum has been described as the pipeline or conveyer belt through which supplies are moved about the factory (or the cell). Viewed through a microscope, the endoplasmic reticulum looks like flat, tubular membranes that are connected to the nucleus and which spread throughout the interior of the cell. There are two types of endoplasmic reticulum: rough endoplasmic reticulum and smooth endoplasmic reticulum. Rough endoplasmic reticulum appears to have a pockmarked, or rough outer surface, and this is because it is covered with ribosomes, which are structures in the cell that make proteins. Smooth endoplasmic reticulum has no ribosomes dotting its surface. Smooth endoplasmic reticulum has the job of packaging and distributing materials needed for making proteins and other substances, and also is responsible for "detoxifying" any substance in the cell that might be poisonous. Both types of endoplasmic reticulum are essential to the cell if it is to be able to move materials around to where they are needed. There is always a large amount of endoplasmic reticulum within a cell, and in many cells it takes up nearly half the amount of interior space. Although both are involved with transport, smooth endoplasmic reticulum curves around the cell's cytoplasm (its jelly-like fluid) like interconnecting pipelines, while rough endoplasmic reticulum looks more like stacked and flattened sacs. Both smooth and rough endoplasmic reticulum are connected directly to many other organelles in the cell, and without the membrane pipeline that the endoplasmic reticulum provides, eukaryotic cells would not be able to carry out many of their functions.

[*See also* **Cell; Organelle**]

Entomology

Entomology is the branch of zoology dealing with insects. It includes the study of the development, anatomy (structure), physiology (functions), behavior, classification, genetics, and ecology of insects. It is a fascinating branch of science because by their sheer numbers, insects are considered to be the dominant group of animals on Earth today.

Based on the Greek word *entomon* for insects, entomology is the scientific study of insects. As with most branches of zoology, entomology was first seriously studied in the work of the Greek philosopher Aristotle (384–322 B.C.). It was he who gave the first good descriptions of insect anatomy and laid the groundwork for entomology by stating that all insects have a body divided into three parts. After Aristotle's work, there was little attention paid to insects except to add more types to the list of known species. However, in 1602, the Italian naturalist Ulisse Aldrovandi (1522–1605) devoted an entire book to insects called *Of Insect Animals,* and by the time the microscope was invented in the latter part of that century, two men in different countries made the first accurate studies of insect anatomy. In Italy, the physiologist Marcello Malpighi (1628–1694) turned his microscope on insects and discovered the tiny, branching tubes with which they breathe. He also devoted an entire volume to the interior

JAN SWAMMERDAM

Dutch naturalist Jan Swammerdam (1637–1680) is considered to be the founder of entomology or the study of insects. He also was the first to observe red blood corpuscles (an unattached body cell) and did a great deal of work in comparative anatomy (the study of the structure of living things).

Swammerdam's life always seemed to be characterized by extremes. Although he led a productive life, it also was extremely difficult and sad. Born in Amsterdam in the Netherlands, the son of an apothecary, or pharmacist, Swammerdam was interested in natural history and particularly insects at a very early age. Although his father sent him to medical school, he intended that his son become a priest. When the young Swammerdam graduated from the University of Leiden, he took up the study of natural history and never practiced medicine. Since he had also refused to become a priest, his father decided not to support him and cut him off from any financial help. Despite a lack of money, Swammerdam continued to pursue his biological studies and did an enormous amount of work under very difficult conditions.

Swammerdam studied the life cycles and anatomy of many species of insects, particularly honeybees, mayflies, and dragonflies. Using his microscope, he made several discoveries about what goes on when certain insects undergo complete metamorphosis. This is when it experiences a complete change in its body shape, such as when a caterpillar turns into a butterfly. From these observations of metamorphic development, Swammerdam was able to classify insects into four major groups, three of which are accepted and used today in insect classification. Aside from this, one of his most significant achievements may have been simply to disprove some of the false notions about insects.

Until Swammerdam demonstrated that insect bodies have real structure and internal organs, most had considered them to have simply fluid-filled

organs of the silkworm. About the same time in Holland, the Dutch naturalist Jan Swammerdam (1637–1680) was doing such excellent and highly detailed studies on insect anatomy, that he came to be considered the founder of modern entomology.

Today's entomology deals with a group of animals that are by far the largest of all the classes. There are more than 800,000 known species of insects, and it is estimated that there may be an equal number of still-unknown species. Insects belong to the phylum Arthropoda, which is considered the most successful group of animals on Earth. This is because insects have lived on Earth for about 350,000,000 years, while humans

cavities with no specialized organs. Since virtually every habitat on land contains insects, and they make up about three-quarters of all the animal species on Earth, Swammerdam was finally giving nature's most prolific and most successful class of animals their due. For example, his detailed drawings and descriptions of their reproductive organs showed that they reproduced sexually rather than via such ancient myths as spontaneous generation (which said that living things can be generated from nonliving things).

Swammerdam also used his microscope to study some of the internal systems of vertebrates (animals with a backbone). In 1658, he documented certain small particles that he observed in the blood of a frog, making this the first time that anyone had ever seen red blood cells. He discovered valves in the lymph system (a network of vessels that carry lymph throughout the body) of mammals and also studied the fertilization of eggs as well as their development into an embryo. Experimentally, he showed that the muscle removed from a frog's leg can be stimulated and made to contract. He also theorized and made an excellent guess about the role of oxygen in respiration. He even pioneered the practice of injecting dyes into a cadaver (a dead body) in order to better observe certain anatomical details. Unfortunately, Swammerdam never published his work and most of it went unknown during his lifetime. Since he had little or no income, he did without a great deal and actually suffered both physically and mentally. Eventually he became sick and undernourished, and with overwork and worry, he soon became depressed and mentally unstable. In 1673 he came under the influence of a cultlike figure to whom he remained devoted until his death at the age of forty-three. It was not until 1737 that the Dutch physician Hermann Boerhaave (1668-1738), discovered Swammerdam's work and paid to have it published. Titled the *Bible of Nature,* this two-volume Latin translation of his Dutch work contains some of the finest illustrations of insects ever produced and served to lay the foundation for entomology.

have been on Earth less than 2,000,000 years. While insects are considered pests by many people, they are in fact highly beneficial to humans in a number of ways. Insects perform many important functions in any given ecosystem (an area in which living things interact with each other and the environment). Many insects are soil-dwellers, wood-borers, and consumers of dead animals, and therefore help in the decomposition process (breaking things down) and in the recycling of nutrients. Many insects are eaten by fish and birds and are thus an important food source. Many insects kill and eat other insects, helping to regulate populations. Many flowering plants are dependent on the pollinating activities of bees, butterflies, and flies in order to reproduce. Insects also provide people with a number of useful products such as silk, wax, and honey. On the other hand, insects can be terribly harmful by causing huge losses of food and acting as transmitters of diseases to humans, plants, and animals.

Modern entomology has two aspects: the scientific side that simply wants to learn as much as possible about all insects, and the practical or applied side (often called economic entomology) that searches for methods to better control insects. This latter part of entomology investigates the physiology, development, genetics, diseases, and behavior of pest insects in order to discover new ways of controlling insect populations. For example, research on insect development has led to the use of specific chemicals that disrupt certain hormones important during metamorphosis (as when a caterpillar changes into a flying insect). Other investigations have shown ways to use chemicals to modify insects' behavior (and perhaps confuse them during their mating cycle). Such methods, when combined with traditional chemical pesticides, come under the term pest management. However, entomologists do not want to exterminate all insects. Rather, they hope to control and limit any bad effects insects may have. This is probably wise, since most people who study insects realize there are many good reasons why insects are beneficial to humans, and why insects are the most abundant and most successful animal group ever.

A mosquito fossilized in amber. Insects are extremely helpful to scientists in learning what Earth was like millions of years ago. (Reproduced by permission of JLM Visuals.)

Insects can also be enormously interesting. They have a skeleton on the outside of their bodies; they have no lungs; they smell with their antennae; and some taste with their feet. Some hear with special organs in their abdomens or front legs. Some are highly social and live in colonies. Because insects are highly adaptable, many people say that only insects will remain after every other living thing has disappeared from Earth.

[*See also* **Insects**]

Enzyme

An enzyme is a protein that acts as a catalyst (a substance that speeds up a chemical reaction) and speeds up chemical reactions in living things. The chemistry of life would not be possible without enzymes since they allow reactions in organisms to happen very quickly and therefore support life. Each enzyme is highly specific and will only work in one particular reaction.

Like a catalyst, an enzyme is not consumed or used up during a reaction. Since it is not changed or affected in any way by the process it helps create, an enzyme is immediately ready to be used again for the same purpose. Enzymes are essential to living things because without them, most of the chemical reactions that take place inside an organism would happen too slowly to keep it alive. Temperature is a key factor in a chemical reaction, and the temperature of most living cells is too low to allow the necessary reactions to take place quickly enough. With the proper enzyme, temperature no longer is a limiting factor. Since without the right enzyme, the reaction might occur so slowly that the cell would die. Although enzymes are catalysts, they are far from being typical—while a catalyst can be any type of simple substance, an enzyme is a complex protein. Further, it is by far the most efficient catalyst, since it can, at times, increase reaction rates by factors of 1,000,000 or more.

Since enzymes are complex proteins, they have their own unique three-dimensional shape. It is this shape that makes a particular enzyme act in a certain way. Their shape is the determining factor in how they will act because of the nature of a chemical reaction. A chemical reaction results in the formation of a new compound from existing ones. The mechanism of a chemical reaction involves either the breaking or the forming of a chemical bond (the link between its atoms). Whether breaking or forming bonds, energy is always necessary, and it is here that enzymes play their part. Enzymes either add energy to make something happen or reduce the energy required for something to happen. Either way,

ANSELME PAYEN

French chemist Anselme Payen (1795–1871) investigated the chemical reactions carried out by plants and discovered diastase, the first enzyme known to science. He also introduced the filtering properties of activated charcoal and discovered cellulose, a basic constituent of plant cells.

Anselme Payen was born in Paris, France, the son of a lawyer who turned to a career in industry and started up several chemical production factories. At the age of twenty, Payen was put in charge of his father's borax production plant. Borax is a crystalline mineral often found in salt lakes that is refined and used in metallurgy (metal-making) and to make soaps, glass, and pottery glazes, among other things. Payen discovered a method of preparing borax from boric acid which was readily available from Italy. Since his production costs for this method were so low, he was able to undersell his competitors, the Dutch, who had a monopoly on borax. Five years later he turned his attention to his company's production of sugar from sugar beets. Seeking a way to remove color impurities from beet sugar, he invented a method of using activated charcoal to filter out, or catch, large molecules. As a result of Payen's process, the filtering properties of charcoal have since been put to many uses, the most notable of which was their use in gas masks during wartime.

In 1833, Payen made another major discovery. That year he separated a substance from malt extract (grain that was germinating or starting to sprout) that seemed to be responsible for speeding up the conversion of starch into sugar. Further tests suggested to Payen that this substance acted as yeast did, meaning that it acted as a catalyst. A catalyst is a chemical that

they are able to create a reaction that would not have occurred without the enzyme.

Since each enzyme has a unique shape, it will only "fit" or work for one particular reaction. This means the more reactions an organism needs, the greater the number of different enzymes required. For example, in the process called cellular respiration (in which food is broken down to release energy), approximately thirty chemical reactions take place, and each is controlled by its own enzyme. It is estimated that the typical animal cell (roughly one-billionth the size of a drop of water) contains about three thousand different enzymes, all of which are programmed to work in a certain chemical reaction.

Enzymes work at the cellular level and can be considered a cell's chemical regulator or system of control. The needs of a living cell are

usually speeds up the rate of a chemical reaction, but is not itself affected, or changed, in any way. Payen named this new substance "diastase" from a Greek word meaning "to separate," since in many ways it separated the individual building blocks of starch into its individual components of sugar. To his delight, he found that diastase even worked when it was taken out of the original malt extract that produced it. Diastase, therefore, could be called an organic catalyst, or an enzyme, since an enzyme is a substance that acts as a catalyst in biological systems. Thus, diastase was the first enzyme to be produced in concentrated form. Ever since, enzymes that have been discovered have been named with the "-ase" suffix, the pattern started by Payen.

Payen was the first to isolate cellulose, a carbohydrate (a compound consisting of only carbon, hydrogen, and oxygen). While studying different types of wood, he obtained a substance that was definitely not starch but that nonetheless could be broken down into units of sugar as starch could. Because he obtained it from the cell walls of plants, he named it "cellulose." Once more, Payen established a naming system, and ever since carbohydrates always end with "-ose," like glucose and sucrose. Much later, cellulose went on to be used as the building block for many other products. Treated with acids and other additives, it was the main ingredient in the manufacture of guncotton (an explosive), celluloid (film), cellophane, and rayon, among others. By 1835 Payen abandoned business altogether and became professor of industrial and agricultural chemistry in Paris. He died in Paris during the Franco-Prussian War after refusing to leave the city as the Prussian army advanced.

constantly changing as it strives to adjust to the ever-changing demands of its environment. Therefore, the cell needs the flexible system of control that enzymes provide. Another advantage of an enzyme is that the reactions that it stimulates are reversible if the cell wants the opposite reaction.

The synthesis, or the making, of an enzyme is controlled by a specific gene. Usually it is a hormone that switches on the gene, which, in turn, signals that a certain enzyme should be produced. Since enzyme production can be turned on, it can also be turned off. Both of these on/off mechanisms are natural and happen all the time, but certain forms of artificial "inhibitors" can turn off production permanently. Many poisons and some drugs have this effect, and can lead to death. For example, certain nerve gases as well as the poison cyanide permanently inhibit or stop the important enzyme that allows the body to use oxygen. This results in a quick death. Besides the natural enzymes in the bodies of organisms, enzymes are put to use everyday in the production of beer, wine, cheese, and bread. Without the proper enzymes contained in yeast cells, none of these important food products could be made.

[*See also* **Protein**]

Eutrophication

Eutrophication is a natural process that occurs in an aging lake or pond as it gradually builds up its concentration of plant nutrients. Cultural or artificial eutrophication occurs when human activity introduces increased amounts of these nutrients. These speed up plant growth and eventually choke the lake of all of its animal life.

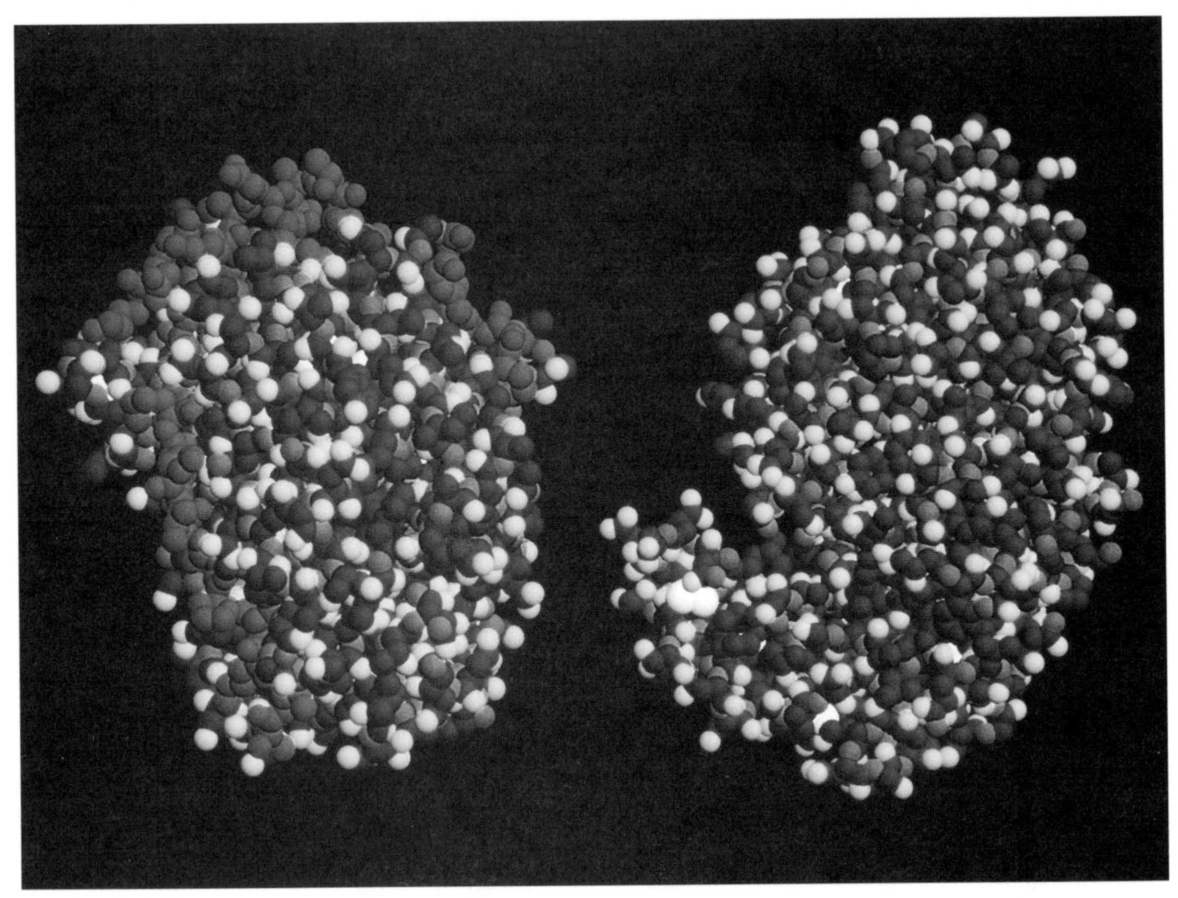

Computer-generated models of pepsinogen, a pre-enzyme found in the stomach, and pepsin, a digestive enzyme. In the presence of increased acidity, pepsinogen transforms into pepsin. (©Ken Edward/Science Source/National Audubon Society Collection/Photo Researchers, Inc. Reproduced by permission.)

In nature, eutrophication is part of the normal aging process of many lakes and ponds. Some never experience it because of a lack of warmth and light, but many do. Over time, these bodies of fresh water change in terms of how productive or fertile they are. While this is different for each lake or pond, those that are fed rich nutrients from a stream or river or some other natural source are described as "eutrophic," meaning they are nutrient-rich and therefore abundant in plant and animal life. Eutrophication is not necessarily harmful or bad, and the word itself is often translated from the Greek as meaning "well nourished" or "good food." However, eutrophication can be speeded up artificially, and then the lake and its inhabitants eventually suffer from too much of a good thing.

HUMANS INCREASE RATE OF EUTROPHICATION

Artificial or cultural eutrophication has become so common, that the word eutrophication by itself has come to mean this harmful acceleration of nutrients. Human activities almost always result in the creation of waste, and many of these waste products often contain nitrates and phosphates. Nitrates are a compound of nitrogen, and most are produced by bacteria. Phosphates are phosphorous compounds. Both nitrates and phosphates are absorbed by plants and are needed for growth. However, the human use of detergents and chemical fertilizers has greatly increased the amount of nitrates and phosphates that are washed into our lakes and ponds. When this occurs in a sufficient quantity, they act like fertilizer for plants and algae and speed up their rate of growth.

Algae are a group of plantlike organisms that live in water and can make their own food by photosynthesis (using sunlight to make food from simple chemicals). As plants begin to grow explosively and algae "blooms," two harmful things occur, both involving oxygen. First, the fast-growing plants and algae consume more oxygen than usual. Algae consume oxygen even at night. This in itself is not that harmful, but when combined with the second effect, it sometimes has a fatal result on the body of water. All algae eventually die, and when they do, oxygen is required by bacteria in order for them to decompose this dead material and break it down. A cycle then begins in which more bacteria decompose more dead algae (consuming more oxygen). The bacteria then release more phosphates back into the water. These phosphates feed more algae.

Eventually, the lake or pond begins to fill in and starts to be choked with plant growth. As the plants die and turn to sediment that sinks, the lake bottom starts to rise. The waters grow shallower and finally the body of water is filled completely and disappears. This can also happen to wetlands, which are already shallow. Eventually, there are shrubs growing where a body of water used to be.

In the 1960s and 1970s, Lake Erie was the most publicized example of eutrophication. Called a "dead lake," the smallest and shallowest of the five Great Lakes was swamped for decades with nutrients from heavily developed agricultural and urban lands. As a result, plant and algae growth choked out most other species living in the lake, and left the beaches unusable due to the smell of decaying algae that washed up on the shores. New pollution controls for sewage treatment plants and agricultural methods by the United States and Canada led to drastic reductions in the amount of nutrients entering the lake. Forty years later, while still not totally free of pollutants and nutrients, Lake Erie is again a biologically thriving lake.

[*See also* **Algae; Bacteria; Pollution**]

A labeled illustration of the structure of an eutrophic lake. (Illustration by Hans & Cassidy. Courtesy of Gale Research.)

Evolution

Evolution is the process by which gradual genetic change occurs over time to a group of living things. As one of the major theories in the life sciences, it can be applied in a narrow sense to the development of a species that adapts to its environment over many generations. In a broader sense, evolution can mean that all life on Earth has progressed from simple to more complex organisms. Evolution is one of the great explanatory theories of science because it describes how biological change occurs over time. However, since it also applies to the human species, it has philosophical and even religious implications and can be considered controversial.

DARWIN PROPOSES THEORY

The theory of evolution was first presented in 1858 by the English naturalist Charles Darwin (1809–1882). Its basic idea is that species undergo genetic change (changes in inherited characteristics) over time and appear different from their long-dead ancestors (although not noticeably different from other living members of their species). Darwin stated that these changes are brought about by the organism's response to the environment. He believed that organisms that are best adapted to their environment have a greater chance of passing on those "fit" traits to their offspring. Nature therefore "selects" the best adapted (fittest) and "selects out" those organisms whose genes contain traits that are less fit.

Darwin's theory further stated that evolution has given rise to all the different forms of life that have ever existed. This means that all species came from a single form of life or a common ancestor that may have first appeared about 3,500,000,000 years ago. The branching out from this first life form that eventually resulted in the enormous variety of living things is called "speciation." Darwin also theorized that the process of evolution is extremely slow. This notion of "gradualism" means that many generations must pass before any changes are noticeable. While Darwin's theory of evolution is not accepted by some who believe in the biblical account of creation, his theory is supported by a vast amount of factual evidence and has withstood all serious attempts to disprove it.

NATURAL SELECTION

The main idea behind evolution is "natural selection," because this concept explains how evolution occurs. Natural selection favors the passing on of beneficial or "fit" genes to the offspring and discourages the transmittal of less valuable genes. As the "engine that drives" evolution, natural selection does not work on the individual unit of life but at the population level. The term population refers to all members of the same species, which are slightly different from one another and live together in a particular place. For example, some deer have larger antlers than others, others are larger and stronger, and others may have a slightly different coloring. Therefore among any population or group of organisms that breed with one another, there will necessarily be a large genetic pool of slightly different traits. Nature, or the population's local environmental conditions, selects which individuals are best adapted. Those individuals with traits that best fit their habitat will get more food, be better prepared to escape a predator, or better able to resist disease than those not so well adapted. Those with favorable traits get to survive and to reproduce—passing these traits on to their young. Individuals with less favored traits do not reproduce as much, and eventually, an entire population will have the traits (in their genes) of the favored group.

Understanding natural selection allows us to understand the basics of evolution. Often the term "survival of the fittest" is used to mean that only the strongest survive or get to live a very long time. Actually, however, the term means that organisms poorly suited or badly "fitted" to their environment are not able to reproduce as much as the "fittest" among them. Consequently, the "fittest" get to survive longer and produce the most offspring. This process assures that it is their genes that get spread the most throughout the population. The fittest organism does not have to be the strongest (or even smartest) individual. What determines fitness is based entirely on the population's environmental conditions. This means that it is essentially a random process. Nature has no master plan that it uses to make species change. Rather, a variety of traits exist in a given population, and whichever one gives an individual an edge over others in the population is the trait that is favored. In many ways, natural selection appears to be a matter of chance. An example of this is the peppered moth of England. Before the Industrial Revolution of the eighteenth century took place and created the steam-powered factories of London, this particular moth was a beige color that enabled it to blend in almost perfectly against local tree bark. Over time, however, as its environment became more and more sooty from the burning of coal, the moth became a much darker color. Its environment change meant that those slightly darker moths had a greater chance of surviving than the lighter ones who were easily seen (and eaten by birds) against a blacker environment.

GENETIC VARIETY

The idea of natural selection is always connected to that of reproduction. It does no good for an organism to live a very long time if it does not reproduce. The fittest not only survive—they reproduce. Of the many factors that influence natural selection, the factor called "genetic variety" is probably the most important. Genetic variety means that the individuals making up a population must have basic genetic differences between them—that is, they have inherited characteristics that make them slightly different from one another. This becomes obvious when we realize that if each individual were identical genetically to all the others, it would not matter which one reproduced. When genetic differences exist, however, who gets to pass on which traits makes a big difference.

Mutations. Genetic variety occurs in two ways: by mutations, which are accidental changes in a gene, and by genetic recombination, which is the mixing of genes that happens to organisms that reproduce sexually. The word mutation often implies something bad, but such is not always the case. Mutations at the genetic level occur during the reshuffling that happens when meiosis takes place. Meiosis (may-OH-sihs) is the type of cell division that makes sex cells. During this division process, genes are sometimes accidentally altered, simply by chance, or by a change in the shape or number of chromosomes, which are the structures that contain the cell's deoxyribonucleic acid (DNA). When a mutation occurs, it means that a particular trait has been changed. Sometimes the mutation is unfavorable since it may eliminate a characteristic that proved beneficial, but it can also be advantageous since it may give a new and even better trait to an individual. If a mutation is favored by natural selection, it gets passed on from one generation to the next. In many ways, pure chance allows mutations to happen and contributes to evolution.

Genetic Recombination. Sexual reproduction is also a source of genetic variety. Sex cells are unlike cells in the rest of the body in that they have only a single set of genes (or only one copy of each gene). It is only after fertilization occurs (when the egg and sperm unite) that the cell gets two copies of each set of genes. During sexual reproduction (the joining of egg and sperm), genes are combined in new ways that result in an individual having a unique combination of genes. This mixing of genes is called genetic recombination, and it accounts for most of the variety needed for natural selection to occur.

ADAPTATION

The end result of natural selection is a process called adaptation. Through natural selection, which favors organisms that "fit" their environment best and weeds out those badly "fitted," living things become better suited, or "adapted," to their habitat or local environment. Darwin called this process of "molding" organisms to their environment over many generations "adaptation." It has come to mean any trait or characteristic that allows an organism to fit better in its environment. Furthermore, it includes an organism's behavior as well as its anatomy (structure) and physiology (internal processes).

THE RATE OF EVOLUTION

Ever since Darwin, the pace or rate at which evolution occurs has been thought to be a gradual one, taking place over a long period of time. However, in recent times two biologists, Stephen Jay Gould of Harvard University and Niles Eldredge of the American Museum of Natural History, have proposed the idea of "punctuated equilibrium." This new hypothesis states that a large number of new species can appear in sudden spurts rather than very gradually over extremely long periods. Gould and Eldredge support their argument by referring to the fossil record, which often shows how some species remained unchanged for millions of years. If change was slow but ever-present, it should appear that way in that fossil record. However, Gould and Eldredge point out cases where new species seem to appear out of nowhere, and use these examples to argue for evolution by sudden bursts.

While there may still be controversy over the rate at which evolution occurs, there is little argument about its major points: living things are constantly competing with one another for the necessities of life; the characteristics of each vary a great deal and are inherited; some of these variations make certain individuals better adapted to their environment; the process of natural selection decides which are more "fit"; more "fit" individuals are more likely to survive and pass on their genes; and all of today's life is the result of evolution progressing from the simple to the complex.

[*See also* **Adaptation; Evolution, Evidence of; Human Evolution; Evolutionary Theory; Natural Selection**]

Evolution, Evidence of

As a scientific theory stating that species undergo genetic change over time and that all living things originated from simple organisms, evolution is considered the most important basic concept in the life sciences. The theory has a great deal of explanatory power, and scientists from many different fields have accumulated a large body of evidence to support it.

DARWIN LAYS FOUNDATION FOR EVOLUTION THEORY

Much of the credit for the acceptance of the theory of evolution should go to English naturalist Charles Darwin (1808–1882). Darwin spent a lifetime gathering data and documenting examples that would explain his theory in detail. Although today's scientists may understand how evolution works in a deeper way than their counterparts did in the past, the foundation of their understanding is built upon Darwin's original observations. Darwin's evidence for evolution falls into several categories. The first of these is called adaptive radiation and is based on Darwin's observation that although mammals are found on all the major landmasses of the Earth, the same mammals are not found in the same habitats. Darwin speculated that when continents moved apart, animals became separated and, like Darwin's finches in the Galapagos Islands, they adapted or went their own way in order to best fit into their environment. Darwin found thirteen species of finches on the Galapagos, and later realized that they all evolved from the same ancestor who came there from South America. He also realized that although the original ancestor probably ate seeds, its descendants adapted their original diet over time to better fit their new environment. Some therefore, became insect eaters while others ate seeds of different sizes. This process is known as adaptive radiation—meaning that many different forms have evolved from the same stock.

DIVERGENT EVOLUTION

The fact that widely separated organisms can have common ancestors suggests that, although these organisms have changed quite a bit over time, some of their basic structure must still be the same. It follows that the more structures these organisms have in common, the more closely related they must be. The study of these structural similarities is called comparative anatomy, and when scientists find two different structures that perform entirely different functions but are basically similar, they argue that this provides evidence of what is called divergent evolution. A good example of divergent evolution are the front limbs of three vertebrates (animals with a backbone). When the arm of a human, the wing of a bird, and the flipper of a whale are compared, it becomes apparent that all are made of the same bones and have five digits (finger-like projections) at their ends. However, each limb has adapted to a different way of life in a different environment.

EVIDENCE OF EVOLUTION

Evidence for evolution can be found by studying three different areas. These are embryology, the fossil record, and the case of the peppered moth of England.

Embryology. Embryology studies the way organisms develop before they are born. Often embryology provides evidence of a common ancestor that cannot be found by studying the adult organism. A good example are two invertebrates (organisms without a backbone): the ragworm and the marine snail. As adults, they are not at all alike, yet when studied as developing larvae, they are remarkably similar and in fact, share a common ancestor.

The Fossil Record. Another area that provides some of the strongest evidence for evolution is the fossil record. Paleontology (the study of fossils) clearly shows that there are many species that no longer exist. By using certain techniques such as carbon dating (a method of determining the age of fossils by measuring the amount of carbon in them) and by studying the placement of fossils within the ground, many fossils can be dated or their age closely estimated. If the fossil record is then placed together based on the ages of each, a gradual change in the form of an organism can be followed and traced eventually to species that exist today. One of the best fossil records of this type is the gradual evolution of reptiles into mammals. Although there are many gaps in the overall fossil record, enough data has been discovered to allow paleontologists to piece together a progression from one-celled life, to simple, multicelled descendants, to more and more complex organisms. Most agree that the weight of fossil evidence in favor of evolution is overwhelming.

The Case of the Peppered Moths. Probably the most direct evidence for evolution and also best known is the case of the peppered moths of England. Prior to the Industrial Revolution of the eighteenth century and the development of the factory system, this species of moth blended perfectly with the pale color of lichens (organisms made up of both fungi and algae) found on London trees and were able to avoid being spotted by hungry birds. Decades after the factory system began, pollution from the factories had killed the lichen and blackened the tree trunks, giving the few dark moths that were part of the species an advantage, since now they blended better with their environment. Eventually, dark peppered moths came to be dominant. Finally, when the pollution sources were reduced and eventually removed, the lighter-colored moth began to make a comeback. Direct observation of such a clear-cut and rapid case of evolution in a living species is rare, but very convincing. Today, plant and animal breeders use methods similar to evolution's natural selection to produce new and more desirable varieties. Their ability to bring about dramatic changes in a fairly short time is more than enough evidence for evolution.

[*See also* **Evolution; Evolutionary Theory; Fossil; Human Evolution**]

Evolutionary Theory

In 1858, when English naturalist Charles Darwin (1809–1882) proposed his theory of evolution by natural selection, most of the scientific world believed that the account of the creation of the world as written in the biblical book of Genesis was true. Most thought that the world was not more than a few thousand years old and that every species had been created separately and had remained basically unchanged. However, Darwin's theory would change that thinking. Even though evolutionary processes had been considered before the publication of Darwin's 1858 book, Darwin's theory rocked the world like no scientific idea ever had, before or since.

ANCIENT GREEKS HINT AT EVOLUTION

Before Darwin conceptualized the idea of evolution, pieces of his theory had been considered and written about by historical thinkers. Some think the idea of evolution began in Greece more than 2500 years ago when the Greek philosopher, Anaximander (610–c.546 B.C.), taught that the life on Earth had begun in the water and eventually moved to dry land. Centuries later, the Greek philosopher Aristotle (384–322 B.C.) taught that the structures of living things were determined by the purpose they would fulfill. Both of these ideas are part of the theory of evolution. For the next 150 years however, no one took these ideas any further, and it was not until the eighteenth century that scientists (then called "natural philosophers") began to think seriously about the origin and progress of life on Earth.

CLASSIFICATION SYSTEM SUPPORTS EVOLUTION

A major step was taken in 1735 when the Swedish botanist, Carolus Linnaeus (1707–1778), first published his *System of Nature* in which he offered science his system of classifying living things. This great accomplishment gave science a way of ordering or making sense of the immense variety of organisms in the natural world. The system also enabled scientists to begin to make generalizations (which would lead to hypotheses or educated guesses and then to theories). Linnaeus also made another contribution toward evolutionary thinking when he dared to include human beings in his classification. It was Linnaeus who called people *Homo sapiens,* which means belonging to the genus *Homo* (man) and the species *sapiens* (wise). This is in contrast to another member of the genus *Homo,* the orangutan, which he designated for some reason as *Homo troglodytes* (man, cave-dwelling).

BUFFON TOUCHES ON THE THEORY OF EVOLUTION

By the mid-1700s, the French naturalist Georges Louis Buffon (1707–1788) was beginning to think seriously if not systematically about the possibilities of life on Earth changing over time. Although he never stated any theory of evolution, his study of the fossil record and his knowledge of comparative anatomy led him to write that life on Earth had undergone some major changes over what appeared to be a very long time. Although he could not explain how or why this had happened, his ideas were definitely based on some parts of evolution.

LAMARCK BECOMES FIRST TO ARGUE FOR EVOLUTION

Finally, at the beginning of the nineteenth century, a student of Buffon's named Jean Baptiste Lamarck (1744–1829) became the first major scientist to put forward a serious, scientific argument for evolution. He also was the first to suggest what type of mechanism would account for evolutionary change. Lamarck's theory argued that evolution occurred through the inheritance of acquired characteristics. His theory, however, was incorrect since it was based on two mistakes. First, he stated that nature had a definite plan and that it kept trying to perfect or make a better organism each time a change occurred. Second, he thought that the characteristics an organism acquired during its own lifetime could be passed on directly to the next generation. Lamarck used the recently discovered giraffe as an example, saying that their ancestors, in trying to reach high leaves to eat, would have stretched their necks and passed this trait on to their offspring. Although we now know that from a genetic standpoint this is not possible, Lamarck's brave new ideas brought the idea of evolution to the scientific forefront.

DARWIN PROPOSES THEORY OF EVOLUTION

Although Lamarck was the first scientist to present and argument for evolution, it was actually Darwin who introduced the theory of evolution that is accepted today. In 1858, Darwin finally published his theory in his

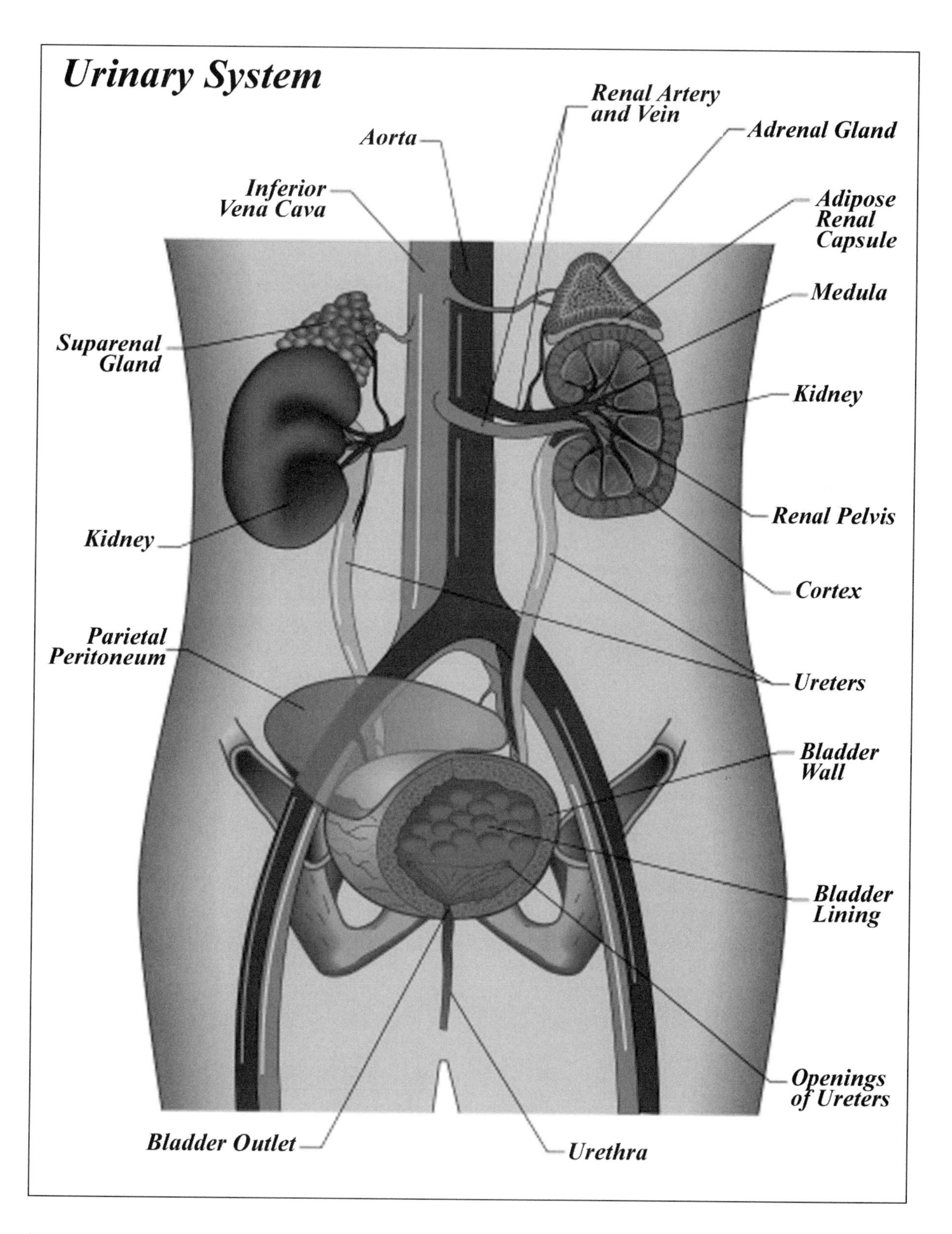
Urinary System
Renal Artery and Vein
Aorta
Adrenal Gland
Inferior Vena Cava
Adipose Renal Capsule
Medula
Suparenal Gland
Kidney
Kidney
Renal Pelvis
Cortex
Parietal Peritoneum
Ureters
Bladder Wall
Bladder Lining
Openings of Ureters
Bladder Outlet
Urethra

comes full (its adult capacity is about two cans of soda), after which it signals to be emptied and urine is expelled through the tube called the urethra.

When kidneys do not work properly, it is difficult to maintain homeostasis and severe disruptions of the body's chemistry can result. Fluid may be retained throughout the body and swelling may occur. Actual kidney failure is a serious, life-threatening condition that can be caused by infection, poisoning, tumors, shock, circulatory disorders, and immune disorders. Treatment for chronic kidney failure includes dialysis, which circulates a patient's blood through a mechanical filtration system outside its body. Permanent kidney failure can only be cured by a transplant.

Extinction

Extinction is the permanent disappearance of an entire species. (A species is a group of closely related, physically similar living organisms that can breed with one another.) A species goes extinct when every single one of its kind has died. Extinctions have been happening since life first began on Earth. However, human activity has greatly accelerated the pace of extinction, and many believe that a great extinction may be taking place today.

While some use the word extinction to describe what might better be described as "local or regional extinction," to most it means the complete and total elimination of a certain species from the face of Earth. Others describe extinction as the loss of a species and its replacement by an evolved version. However, although a particular species may no longer exist, the fact that it adapted or changed so much that it became a distinctly different species means that it never died out completely. Scientists know from fossils that the earliest ancestor of today's horse stood less than 2 feet (0.61 meters) high. Although that species of horse no longer exists, in a way it does, because it slowly and gradually evolved into the very large horse that is common today. If the smaller horse had died out completely and never had a chance to adapt and change and to pass on those changes to its descendants, then it would have gone extinct.

A NATURAL PART OF EARTH'S LIFE CYCLE

Extinction is an ongoing feature of the Earth's ever-changing ecosystem. It is usually caused by major environmental or climate changes. Humans also indirectly cause extinction by the effects their activities have on the environment. Hunting, habitat destruction, and pollution are some of the ways that humans have driven species to extinction. Also, a phenomenon known as "coextinction" can take place when the disappearance of one species of plant or animal dooms the existence of others that are entirely dependent upon it. It is believed that the saber-toothed tiger of prehistoric times died out completely when its sole prey, the mastodon, went extinct.

Throughout the enormously long time that life has existed on Earth, millions of species have simply come and gone. It is estimated that more than 90 percent of all the species that ever existed during Earth's history are now extinct. This is known from the fossil record, which reveals that a number of mass extinctions took place; huge events that resulted in the destruction of large numbers of different species. (A fossil is the remains of a living thing that has tuned into rock and therefore has been preserved.) The different layers, or strata, of Earth's crust represent different times in Earth's history. By this record, we know that the earliest mass extinction was about 650,000,000 years ago during a time called the Precambrian period when life on Earth consisted mostly of algae (plants or plantlike organisms that contain chlorophyll and other pigments that trap light from the Sun) floating on oceans. Scientists think that about 70 percent of the algae species were killed by a major ice age (a period where glaciers covered much of the Earth). Another mass extinction, probably caused by a climate change and considered the largest ever, occurred about 250,000,000 years ago, killing about 90 percent of all land and sea creatures. However, the best-known extinction killed the dinosaurs and many other species during the Cretaceous period about 65,000,000 years ago. Scientists are now fairly certain that this event was caused by a meteorite that crashed into the Earth, sending a vast cloud of dust into the atmosphere and blotting out all sunlight. Most think that the food chain was disrupted since plants could not grow without light, resulting in the death of herbivores (plant-eaters) and the eventual death of carnivores (flesh-eaters). Scientists do not agree as to whether the dense cloud would have cooled the atmosphere or warmed it, but either scenario would have changed the environment so quickly that plant and animal life would have had no time to adapt. Finally, the Ice Age that occurred during the Pleistocene era (between 1,600,000 and 10,000 years ago) caused the destruction of many plants and animals. This was a time of extremely diverse animal life, and mammals rather than reptiles were the dominant species.

HUMANS INCREASE EXTINCTION RATE

Since the evolution (the process by which gradual genetic change occurs) of humans about 10,000 years ago, there have been dramatic changes

LUIS WALTER ALVAREZ

American physicist (a person specializing in the study of physics) Luis Alvarez (1911-1988) was a physicist who had wide-ranging interests and abilities that led him to suggest that the extinction of dinosaurs was caused indirectly by an asteroid that struck Earth. During World War II (1939-45), he was involved with radar and the atomic bomb, and in 1968 he won the Nobel Prize for Physics.

Luis Walter Alvarez was born in San Francisco, California, the son of a medical researcher and physician. His paternal grandfather was originally from Spain, but had run away to Cuba before making his fortune in real estate in California. Young Alvarez attended school in San Francisco, but when his father accepted a position at the Mayo Clinic in Rochester, Minnesota, he attended high school there. After enrolling in the University of Chicago in 1928 to study chemistry, he soon came to love physics and switched his major. Alvarez stayed at Chicago through his bachelor's, master's, and the doctorate degree he received in 1936. He then joined the faculty at the University of California at Berkeley, where he remained until his retirement in 1978.

At Berkeley, Alvarez soon was given the title "prize wild idea man" by his colleagues because of his involvement in such a wide variety of research activities. His earliest research was in the area of nuclear physics and cosmic rays, and when World War II broke out in Europe he worked at Massachusetts Institute of Technology's radiation laboratory on radar (a method of

in Earth's biosphere (all the parts of Earth that make up the living world). With the invention of agriculture (farming) and the domestication (taming) of some animals, humans spread throughout the world and began to use its resources. Some ecologists argue that early humans hunted certain species into extinction, but probably more have been simply crowded out by countless human populations who have taken over the species' habitat. For example, it is known that at least 200 plant species that were native to North America have become extinct only in the past few thousand years. Worldwide, it is estimated that as many as 20 percent of all bird species have disappeared since the evolution of humans.

While the threat of extinction has always existed in nature, the actual time it takes to happen has rapidly speeded up in modern times. The difference today is the overwhelming pressure placed upon the environment by human activity. Habitat destruction, pollution, and over-harvesting are some of the reasons why today, some ecologists argue that we are driving species to extinction at a rate between 1,000 and 10,000 times faster

detecting distinct objects) and helped develop a narrow beam radar system that allowed airplanes to land in bad weather. He was also involved in the Manhattan Project to develop the world's first nuclear weapons. Since he helped develop the bomb's detonating device, Alvarez flew aboard a B-29 airplane that followed the *Enola Gay* when it dropped the first atomic bomb on Hiroshima, Japan. Returning to Berkeley after the war, Alvarez built a huge bubble chamber that could track extremely short-lived particles. It was for this invention that he won the Nobel Prize for Physics in 1968.

In 1980, he and his son, Walter, who was a professor of geology at Berkeley, accidentally discovered a band of sedimentary rock in Italy that contained an unusually high level of the rare metal iridium. Dating techniques set the age of this layer at about 65,000,000 years old. Similar iridium-rich layers were later found in other parts of the world. This led the father-son team to propose a theory regarding the extinction of dinosaurs that occurred 65,000,000 years ago. They then proposed that the iridium had come from an asteroid that struck Earth and sent huge volumes of smoke and dust (including the iridium) into Earth's atmosphere. They suggested that the cloud produced by the asteroid's impact covered Earth for so long that it blocked out sunlight, causing widespread death of plant life. This loss of plant life in turn brought about the extinction of dinosaurs who fed on the plants. While this overall theory has found favor among many scientists and has been confirmed to some extent by additional findings, it is still the subject of much debate, although most now agree with the first half of the hypothesis (the asteroid impact).

than has ever happened before. The American biologist Edward O. Wilson (1929–) currently estimates that the world may have already lost one-fifth of its present species. Today's human interference in nature is probably the most destructive since much of it is the result of chemical pollutants that easily enter an organism's reproductive system and cause a wide range of birth defects. Habitat destruction also accounts for a great amount of loss as the rain forests of the world, which contains a vast variety of plant and animals species, are steadily being cut down.

COUNTRIES TRY TO CURB EXTINCTION RATES

Many countries are trying to control the effects of human activity, which can be most devastating but is also most manageable. Laws have been passed in many countries that protect certain habitats from destruction and individual species from being hunted. Zoos are changing from places that exhibit animals to places that keep an endangered species alive. One such success story is the resurgence of the American bison. Where it once dominated the plains from Alaska and western Canada to the United States and northern Mexico, the American bison was so ruthlessly hunted that by 1899 there were fewer than 1,000 left. When guarded preserves were established and breeding programs begun, the numbers of plains bison rose to more than 50,000 and the species is no longer threatened by extinction. The bison resurgence however, required an enormous and expensive effort on a national scale, and would be impossible to duplicate for every valued species that is threatened. Instead, it is easier and more efficient to regulate the activities that threaten the diversity of life on Earth in the first place. While this may sound simple however, it often involves very sensitive issues of economics, which many times takes priority over the protection of plant and animal species.

[*See also* **Endangered Species; Habitat; Pollution**]

For Further Information

Books

Abbot, David, ed. *Biologists.* New York: Peter Bedrick Books, 1983.

Agosta, William. *Bombardier Beetles and Fever Trees.* Reading, Mass.: Addison-Wesley Publishing Co., 1996.

Alexander, Peter and others. *Silver, Burdett & Ginn Life Science.* Morristown, N.J.: Silver, Burdett & Ginn, 1987.

Alexander, R. McNeill, ed. *The Encyclopedia of Animal Biology.* New York: Facts on File, 1987.

Allen, Garland E. *Life Science in the Twentieth Century.* New York: Cambridge University Press, 1979.

Attenborough, David. *The Life of Birds.* Princeton, N.J.: Princeton University Press, 1998.

Attenborough, David. *The Private Life of Plants.* Princeton, N.J.: Princeton University Press, 1995.

Bailey, Jill. *Animal Life: Form and Function in the Animal Kingdom.* New York: Oxford University Press, 1994.

Bockus, H. William. *Life Science Careers.* Altadena, Calf.: Print Place, 1991.

Borell, Merriley. *The Biological Sciences in the Twentieth Century.* New York: Scribner, 1989.

Braun, Ernest. *Living Water.* Palo Alto, Calf.: American West Publishing Co., 1971.

Burnie, David. *Dictionary of Nature.* New York: Dorling Kindersley Inc., 1994.

Burton, Maurice, and Robert Burton, eds. *Marshall Cavendish International Wildlife Encyclopedia.* New York: Marshall Cavendish, 1989.

Coleman, William. *Biology in the Nineteenth Century.* New York: Cambridge University Press, 1977.

Conniff, Richard. *Spineless Wonders.* New York: Henry Holt & Co., 1996.

Corrick, James A. *Recent Revolutions in Biology.* New York: Franklin Watts, 1987.

Curry-Lindahl, Kai. *Wildlife of the Prairies and Plains.* New York: H. N. Abrams, 1981.

Darwin, Charles. *The Origin of Species.* New York: W.W. Norton & Company, Inc., 1975.

Davis, Joel. *Mapping the Code.* New York: John Wiley & Sons, 1990.

Diagram Group Staff. *Life Sciences on File.* New York: Facts on File, 1999.

Dodson, Bert, and Mahlon Hoagland. *The Way Life Works.* New York: Times Books, 1995.

Drlica, Karl. *Understanding DNA and Gene Cloning.* New York: John Wiley & Sons, 1997.

For Further Information

Edwards, Gabrielle I. *Biology the Easy Way.* New York: Barron's, 1990.

Evans, Howard Ensign. *Pioneer Naturalists.* New York: Henry Holt & Sons, 1993.

Farrington, Benjamin. *What Darwin Really Said.* New York: Schocken Books, 1982.

Finlayson, Max, and Michael Moser, eds. *Wetlands.* New York: Facts on File, 1991.

Goodwin, Brian C. *How the Leopard Changed Its Spots: The Evolution of Complexity.* New York: Simon & Schuster, 1996.

Gould, Stephen Jay, ed. *The Book of Life.* New York: W.W. Norton & Company, Inc., 1993.

Greulach, Victor A., and Vincent J. Chiapetta. *Biology: The Science of Life.* Morristown, N.J.: General Learning Press, 1977.

Grolier World Encyclopedia of Endangered Species. 10 vols. Danbury, Conn.: Grolier Educational Corp., 1993.

Gutnik, Martin J. *The Science of Classification: Finding Order Among Living and Nonliving Objects.* New York: Franklin Watts, 1980.

Hall, David O., and K.K. Rao. *Photosynthesis.* New York: Cambridge University Press, 1999.

Hare, Tony. *Animal Fact-File: Head-to-Tail Profiles of More than 100 Mammals.* New York: Facts on File, 1999.

Hare, Tony, ed. *Habitats.* Upper Saddle River, N.J.: Prentice Hall, 1994.

Hawley, R. Scott, and Catherine A. Mori. *The Human Genome: A User's Guide.* San Diego, Calf.: Academic Press, 1999.

Huxley, Anthony Julian. *Green Inheritance.* New York: Four Walls Eight Windows, 1992.

Jacob, François. *Of Flies, Mice, and Men.* Cambridge, Mass.: Harvard University Press, 1998.

Jacobs, Marius. *The Tropical Rain Forest.* New York: Springer-Verlag, 1990.

Johanson, Donald, and Blake Edgar. *From Lucy to Language.* New York: Simon & Schuster, 1996.

Jones, Steve. *The Language of Genes.* New York: Doubleday, 1994.

Kapp, Ronald O. *How to Know Pollen and Spores.* Dubuque, Iowa: W. C. Brown, 1969.

Kordon, Claude. *The Language of the Cell.* New York: McGraw-Hill, 1993.

Lambert, David. *Dinosaur Data Book.* New York: Random House Value Publishing, Inc., 1998.

Leakey, Richard. *The Origin of Humankind.* New York: Basic Books, 1994.

Leakey, Richard, and Roger Lewin. *Origins Reconsidered.* New York: Doubleday, 1992.

Leonard, William H. *Biology: A Community Context.* Cincinnati, Ohio: South-Western Educational Pub., 1998.

Levine, Joseph S., and David Suzuki. *The Secret of Life: Redesigning the Living World.* Boston, Mass.: WGBH Boston, 1993.

Little, Charles E. *The Dying of the Trees.* New York: Viking, 1995.

Lovelock, James. *Healing Gaia.* New York: Harmony Books, 1991.

McGavin, George. *Bugs of the World.* New York: Facts on File, 1993.

McGowan, Chris. *Diatoms to Dinosaurs.* Washington, D.C.: Island Press/Shearwater Books, 1994.

McGowan, Chris. *The Raptor and the Lamb.* New York: Henry Holt & Co., 1997.

McGrath, Kimberley A. *World of Biology.* Detroit, Mich.: The Gale Group, 1999.

Magner, Lois N. *A History of the Life Sciences.* New York: Marcel Dekker, Inc., 1979.

Manning, Richard. *Grassland.* New York: Viking, 1995.

Margulis, Lynn. *Early Life.* Boston, Mass.: Science Books International, 1982.

Margulis, Lynn, and Karlene V. Schwartz. *Five Kingdoms.* New York: W.H. Freeman, 1998.

Margulis, Lynn, and Dorian Sagan. *The Garden of Microbial Delights.* Dubuque, Iowa: Kendall Hunt Publishing Co., 1993.

Marshall, Elizabeth L. *The Human Genome Project.* New York: Franklin Watts, 1996.

Mauseth, James D. *Plant Anatomy.* Menlo Park, Calf.: Benjamin/Cummings Publishing Co., 1988.

Mearns, Barbara. *Audubon to X'antus.* San Diego, Calf.: Academic Press, 1992.

Moore, David M. *Green Planet: The Story of Plant Life on Earth.* New York: Cambridge University Press, 1982.

Morris, Desmond. *Animal Days.* New York: Morrow, 1979.

Morton, Alan G. *History of the Biological Sciences: An Account of the Development of Botany from Ancient Times to the Present Day.* New York: Academic Press, 1981.

Nebel, Bernard J., and Richard T. Wright. *Environmental Science: The Way the World Works.* Upper Saddle River, N.J.: Prentice Hall, 1998.

Nies, Kevin A. *From Priestess to Physician: Biographies of Women Life Scientists.* Los Angeles, Calf.: California Video Institute, 1996.

Norell, Mark, A., Eugene S. Gaffney, and Lowell Dingus. *Discovering Dinosaurs in the American Museum of Natural History.* New York: Knopf, 1995.

O'Daly, Anne, ed. *Encyclopedia of Life Sciences.* 11 vols. Tarrytown, N.Y.: Marshall Cavendish Corp., 1996.

Postgate, John R. *Microbes and Man.* New York: Cambridge University Press, 2000.

Reader's Digest Editors. *Secrets of the Natural World.* Pleasantville, N.Y.: Reader's Digest Association, 1993.

Reaka-Kudla, Marjorie L., Don E. Wilson, and Edward O. Wilson. *Biodiversity II: Understanding and Protecting Our Biological Resources.* Washington, D.C.: Joseph Henry Press, 1997.

Rensberger, Boyce. *Life Itself.* New York: Oxford University Press, 1996.

Rosenthal, Dorothy Botkin. *Environmental Science Activities.* New York: John Wiley & Sons, 1995.

Ross-McDonald, Malcom, and Robert Prescott-Allen. *Man and Nature: Every Living Thing.* Garden City, N.Y.: Doubleday, 1976.

Sayre, Anne. *Rosalind Franklin and DNA.* New York: W.W. Norton & Co., 1975.

Shearer, Benjamin F., and Barbara Smith Shearer. *Notable Women in the Life Sciences: A Biographical Dictionary.* Westport, Conn.: Greenwood Press, 1996.

Shreeve, Tim. *Discovering Ecology.* New York: American Museum of Natural History, 1982.

Singer, Charles Joseph. *A History of Biology to about the Year 1900.* Ames, Iowa: Iowa State University Press, 1989.

Singleton, Paul. *Bacteria in Biology, Biotechnology and Medicine.* New York: John Wiley & Sons, 1999.

Snedden, Robert. *The History of Genetics.* New York: Thomson Learning, 1995.

Stefoff, Rebecca. *Extinction.* New York: Chelsea House, 1992.

Stephenson, Robert, and Roger Browne. *Exploring Variety of Life.* Austin, Tex.: Raintree Steck-Vaughn, 1993.

Sturtevant, Alfred H. *History of Genetics.* New York: Harper & Row, 1965.

Tesar, Jenny E. *Patterns in Nature: An Overview of the Living World.* Woodbridge, Conn.: Blackbirch Press, 1994.

Tocci, Salvatore. *Biology Projects for Young Scientists.* New York: Franklin Watts, 1999.

Tremain, Ruthven. *The Animal's Who's Who.* New York: Scribner, 1982.

Tyler-Whittle, Michael Sidney. *The Plant Hunters.* New York: Lyons & Burford, 1997.

Verschuuren, Gerard M. *Life Scientists.* North Andover, Mass.: Genesis Publishing Co., 1995.

Wade, Nicholas. *The Science Times Book of Fish.* New York: Lyons Press, 1997.

Wade, Nicholas. *The Science Times Book of Mammals.* New York: Lyons Press, 1999.

Walters, Martin. *Innovations in Biology.* Santa Barbara, Calf.: ABC-CLIO, 1999.

Watson, James D. *The Double Helix: A Personal Account of the Discover of the Structure of DNA.* New York: Scribner, 1998.

Wilson, Edward O. *The Diversity of Life.* Cambridge, Mass.: Belknap Press of Harvard University Press, 1992.

Videocassettes

Attenborough, David. *Life on Earth.* 13 episodes. BBC in association with Warner Brothers & Reiner Moritz Productions. Distributor, Films Inc. Chicago, Ill.: 1978. Videocassette.

Attenborough, David. *The Living Planet.* 12 episodes. BBC/Time-Life Films. Distributor, Ambrose Video Publishing, Inc., N.Y. Videocassette.

For Further Information

Web Sites

ALA (American Library Association): Science and Technology: Sites for Children: Biology. http://www.ala.org/parentspage/greatsites/science.html#c (Accessed August 9, 2000).

Anatomy and Science for Kids. http://kidscience.about.com/kids/kidscience/msub53.htm (Accessed August 9, 2000).

ARS (Agricultural Research Service): Sci4Kids. http://www.ars.usda.gov/is/kids/ (Accessed August 9, 2000).

Best Science Links for Kids (Georgia State University). http://www.gsu.edu/~chevkk/kids.html (Accessed August 9, 2000).

Cornell University: Cornell Theory Center Math and Science Gateway. http://www.tc.cornell.edu/Edu/MathSciGateway/ (Accessed August 9, 2000).

Defenders of Wildlife: Kids' Planet. http://www.kidsplanet.org/ (Accessed August 9, 2000).

DLC-ME (Digital Learning Center for Microbiology Ecology). http://commtechlab.msu.edu/sites/dlc-me/index.html (Accessed August 9, 2000).

The Electronic Zoo. http://netvet.wustl.edu/e-zoo.htm (Accessed August 9, 2000).

Explorer: Natural Science. http://explorer.scrtec.org/explorer/explorer-db/browse/static/Natural/\Science/index.html (Accessed August 9, 2000).

Federal Resources for Educational Excellence: Science. http://www.ed.gov/free/s-scienc.html (Accessed August 9, 2000).

Fish Biology Just for Kids: Florida Museum of Natural History. http://www.flmnh.ufl.edu/fish/Kids/kids.htm (Accessed August 9, 2000).

Franklin Institute Online: Science Fairs. http://www.fi.edu/qanda/spotlight1/spotlight1.html (Accessed August 9, 2000).

GO Network: Biology for Kids. http://www.go.com/WebDir/Family/Kids/At_school/Science_and_technology/Biology_for_kids (Accessed August 9, 2000).

Howard Hughes Medical Institute: Cool Science for Curious Kids. http://www.hhmi.org/coolscience/ (Accessed August 9, 2000).

Internet Public Library: Science Fair Project Resource Guide. http://www.ipl.org/youth/projectguide/

Internet School Library Media Center: Life Science for K-12. http://falcon.jmu.edu/~ramseyil/lifescience.htm (Accessed August 9, 2000).

K-12 Education Links for Teachers and Students (Pollock School). http://www.ttl.dsu.edu/hansonwa/k12.htm (Accessed August 9, 2000).

Kapili.com: Biology4Kids! Your Biology Web Site!. http://www.kapili.com/biology4kids/index.html (Accessed August 9, 2000).

Lawrence Livermore National Laboratory: Fun Science for Kids. http://www.llnl.gov/llnl/03education/science-list.html (Accessed August 9, 2000).

LearningVista: Kids Vista: Sciences. http://www.kidsvista.com/Sciences/index.html (Accessed August 9, 2000).

Life Science Lesson Plans: Discovery Channel School. http://school.discovery.com/lessonplans/subjects/lifescience.html (Accessed August 9, 2000).

Life Sciences: Exploratorium's 10 Cool Sites. http://www.exploratorium.edu/learning_studio/cool/life.html (Accessed August 9, 2000).

Lightspan StudyWeb: Science. http://www.studyweb.com/Science/ (Accessed August 9, 2000).

Lycos Zone Kids' Almanac.
http://infoplease.kids.lycos.com/science.html
(Accessed August 9, 2000).

Mr. Biology's High School Bio Web Site.
http://www.sc2000).net/~czaremba/
(Accessed August 9, 2000).

Mr. Warner's Cool Science: Life Links.
http://www3.mwis.net/~science/life.htm
(Accessed August 9, 2000).

Naturespace Science Place.
http://www.naturespace.com/
(Accessed August 9, 2000).

NBII (National Biological Information Infrastructure): Education.
http://www.nbii.gov/education/index.html
(Accessed August 9, 2000).

PBS Kids: Kratts' Creatures.
http://www.pbs.org/kratts/
(Accessed August 9, 2000).

Perry Public Schools: Educational Web Sites: Science Related Sites.
http://scnc.perry.k12.mi.us/edlinks.html#Science
(Accessed August 9, 2000).

QUIA! (Quintessential Instructional Archive) Create Your Own Learning Activities.
http://www.quia.com/
(Accessed August 9, 2000).

Ranger Rick's Kid's Zone: National Wildlife Federation.
http://www.nwf.org/nwf/kids/index.html
(Accessed August 9, 2000).

The Science Spot.
http://www.theramp.net/sciencespot/index.html
(Accessed August 9, 2000).

South Carolina Statewide Systemic Initiative (SC SSI): Internet Resources: Math Science Resources.
http://scssi.scetv.org/mims/ssrch2.htm
(Accessed August 9, 2000).

ThinkQuest: BodyQuest.
http://library.thinkquest.org/10348/
(Accessed August 9, 2000).

United States Department of the Interior Home Page: Kids on the Web.
http://www.doi.gov/kids/
(Accessed August 9, 2000).

USGS (United States Geological Service) Learning Web: Biological Resources.
http://www.nbs.gov/features/education.html
(Accessed August 9, 2000).

Washington University School of Medicine Young Scientist Program.
http://medinfo.wustl.edu/~ysp/
(Accessed August 9, 2000).

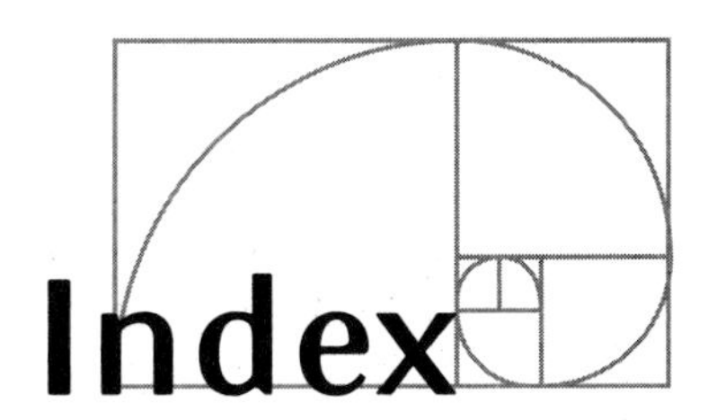

Index

Italic type *indicates volume number;* **boldface** *indicates main entries and their page numbers;* (ill.) *indicates photos and illustrations.*

A

C

D

E

F

G

H

I

J

K

L

M

N

O

P

T

U

V

W

X

Y

Z